The Solution of Equations in Integers

A. O. Gelfond
Moscow University

Translated from Russian and Edited by
J. B. Roberts
Reed College

Dover Publications, Inc.
Mineola, New York

Bibliographical Note

This Dover edition, first published in 2018, is an unabridged republication of the work originally published by W. H. Freeman and Company, San Francisco, California, in 1961.

Library of Congress Cataloging-in-Publication Data

Names: Gelfond, A. O. (Aleksandr Osipovich), 1906–1968, author. | Roberts, Joe, editor.
Title: The solution of equations in integers / A.O. Gelfond, Moscow University ; translated from the Russian and edited by J.B. Roberts, Reed College.
Other titles: Reshenie uravneniæi v ëtiselykh chislakh. English
Description: Dover edition. | Mineola, New York : Dover Publications, Inc., 2018. | Translation originally published: San Francisco, California : W.H. Freeman and Company, 1961.
Identifiers: LCCN 2017047328 | ISBN 9780486824598 | ISBN 0486824594
Subjects: LCSH: Equations—Numerical solutions. | Number theory.
Classification: LCC QA218 .G413 2018 | DDC 512.9/4—dc23
LC record available at https://lccn.loc.gov/2017047328

Manufactured in the United States by LSC Communications
82459401   2018
www.doverpublications.com

FOREWORD

A. O. Gelfond, professor of mathematics at Moscow University, is the author of more than seventy-five books and research papers on mathematics, and is one of the foremost number theorists in the world. His work has had considerable influence on current research in the areas of mathematics discussed here. This book, which is a translation of the 1957 Russian edition, can be read by anyone who is acquainted with high school algebra.

March, 1961   J. B. ROBERTS

PREFACE

This book is based on a lecture on equations in integers which I read to the mathematical Olympiad at the Moscow State University in 1951. I wish here to express thanks to my student, H. M. Korobov, for his help in writing the first, second, and part of the third sections in accordance with an abstract of my lecture. The material discussed in this book is within reach of the upper grades in the high schools.

A. GELFOND

CONTENTS

INTRODUCTION

The theory of numbers covers the basic arithmetical properties of the natural number series, that is, the positive integers,
and belongs to the oldest branch of mathematics. One of the
central problems of the so-called analytic theory of numbers is
concerned with the distribution of the prime numbers in the
natural number series. A prime number is any positive integer
greater than unity and divisible without remainder only by
itself and unity. The problem of the distribution of prime
numbers in the natural number series consists in determining
the regularity of the behavior of the number of prime numbers
less than N for large values of N. The first result in this direc-
tion we find in Euclid (400 B.C.), namely, a proof of the in-
finitude of the series of prime numbers; the second result was
obtained by the famous Russian mathematician P. L. Cheby-
cheff in the second half of the nineteenth century. Another
basic problem in the theory of numbers is the problem of the
representation of integers as the sum of integers of specified
type—for example, representing odd integers as the sum of
three prime integers. The last problem, the problem of Gold-
bach, was solved* comparatively recently by the most sig-
nificant present-day representative of the theory of numbers—
the Soviet mathematician I. M. Vinogradoff.

This book is dedicated to a very interesting branch of the
theory of numbers, the solution of equations in integers. The
determination of integral solutions of algebraic equations with
integral coefficients and with more than one unknown is one

---

* Not strictly correct since what Vinogradoff showed was only that every
sufficiently large integer can be written as the sum of three primes [*Translator*].

of the most difficult problems in number theory. Many of the outstanding mathematicians of antiquity were concerned with this problem—the Greek mathematician Pythagoras (sixth century B.C.), the Alexandrian mathematician Diophantus (third century B.C.)—as well as some of the best mathematicians nearer our time—P. Fermat (seventeenth century), L. Euler and J. Lagrange (eighteenth century), and others. Nevertheless, the efforts of many generations of outstanding mathematicians have failed to yield any general methods comparable to the Vinogradoff method of trigonometric sums which has been used to solve quite distinct problems in the analytic theory of numbers.

The problem of solving equations in integers is completely settled only for equations of second degree in two unknowns. Equations of any degree with one unknown do not present any interest since they can be solved with the help of a finite number of trials. For equations of degree higher than the second with two or more unknowns, it is a very difficult problem not only to find all solutions in integers, but even to answer the simpler question whether a finite or an infinite number of such solutions exists.

Since equations in integers are encountered in physics, the solution of these equations is of more than theoretical interest. However, the theoretical interest in equations in integers is very great since these equations are closely connected with many problems in the theory of numbers. In addition, the elementary parts of the theory of such equations, which are discussed in this book, may be used to widen the horizons both of students and teachers.

In this book are expounded certain fundamental results obtained in the theory of solving equations in integers. The theorms formulated are proved when these proofs are not too difficult.

1. Equations in One Unknown

Let us consider equations of the first degree in one unknown

$$a_1 x + a_0 = 0. \tag{1}$$

Let the coefficients a_1 and a_0 in the equation be integers. It is clear that the solution of this equation,

$$x = - \frac{a_0}{a_1},$$

will be an integer only if a_1 integrally divides a_0. Thus equation *(1)* is not always solvable in integers. For example, of the two equations $3x - 27 = 0$ and $5x + 21 = 0$, the first has the integer solution $x = 9$, but the second is unsolvable in integers.

The same situation is met in equations of degree higher than the first. The quadratic equation $x^2 + x - 2 = 0$ has integer solutions $x_1 = 1$, $x_2 = -2$; but the equation $x^2 - 4x + 2 = 0$ is unsolvable in integers since its roots $x_{1,2} = 2 \pm \sqrt{2}$ are irrational.

The problem of finding integral roots of equations of the nth degree with integral coefficients

$$a_n x^n + a_{n-1} x^{n-1} + \ldots + a_1 x + a_0 = 0 \quad (n \geq 1) \tag{2}$$

is easily solved. Indeed, let $x = a$ be an integral root of this equation. Then

$$a_n a^n + a_{n-1} a^{n-1} + \ldots + a_1 a + a_0 = 0,$$

$$a_0 = - a \, (a_n a^{n-1} + a_{n-1} a^{n-2} + \ldots + a_1).$$

From the last equation it is seen that a_0 is divisible by a without remainder. Consequently, every integer root of equation *(2)* is a divisor of the constant term of the equation. In order to find the integral solutions of the equation, it is necessary to choose those of the divisors of a_0 which, when substituted into the equation, convert it into an identity. Thus, for example, of the numbers 1, -1, 2, -2 representing all divisors of the constant term of the equation

$$x^{10} + x^7 + 2x^3 + 2 = 0,$$

only -1 is a root. Consequently this equation has a unique integral root $x = -1$. In the same way, it is easy to show that the equation

$$x^6 - x^5 + 3x^4 + x^2 - x + 3 = 0$$

has no integral solutions.

Of considerably greater interest is the solution in integers of equations with many unknowns.

2. Equations of the First Degree in Two Unknowns

Let us consider equations of the first degree with two unknowns

$$ax + by + c = 0, \tag{3}$$

where a and b are integers different from 0 and c is an arbitrary integer. We will consider the case that the coefficients a and b do not have a common divisor other than unity.*
Indeed, if the greatest common divisor $d = (a, b)$ is distinct from unity, then $a = a_1 d$, $b = b_1 d$, and equation *(3)* takes the form

$$(a_1 x + b_1 y)d + c = 0.$$

---

* Such numbers a and b are called *relatively prime*; designate by (a, b) the greatest common divisor of the numbers a and b; for relatively prime numbers we have $(a, b) = 1$.

This equation has an integral solution only when c is divisible by d. Thus, in the case $(a, b) = d \neq 1$, all coefficients of the equation *(3)* are necessarily divisible by d, and the canceling of d leads to the equation

$$a_1x + b_1y + c_1 = 0 \qquad (c_1 = c/d),$$

in which the coefficients a_1 and b_1 are relatively prime.

We consider first the case when $c = 0$. Equation *(3)* is then

$$ax + by = 0. \tag{3'}$$

Solving this equation for x yields

$$x = -(b/a)y.$$

It is clear that x will take integral values when and only when a divides y without remainder. But every y which is a multiple of a may be written in the form

$$y = at,$$

where t takes on arbitrary integral values $(t = 0, \pm1, \pm2, \ldots)$. If we substitute these values of y in the preceding equation, then

$$x = -(b/a)at = -bt,$$

and we obtain formulas containing all integral solutions of the equation *(3')*:

$$x = -bt, y = at \qquad (t = 0, \pm1, \pm2, \ldots).$$

We proceed now to the case $c \neq 0$.

We show first that in order to find all solutions of equation *(3)* it is sufficient to find any one of its solutions, that is, to find integers x_0, y_0 for which

$$ax_0 + by_0 + c = 0.$$

THEOREM I. *Let a and b be relatively prime and let* $[x_0, y_0]$ *be any solution* of the equation*

$$ax + by + c = 0. \tag{3}$$

Then the formulas

$$x = x_0 - bt, \quad y = y_0 + at \tag{4}$$

for $t = 0, \pm 1, \pm 2, \ldots,$ *give all solutions of equation (3).*

Proof: Let $[x,y]$ be an arbitrary solution of equation *(3)*. Then from the equalities

$$ax + by + c = 0 \quad \text{and} \quad ax_0 + by_0 + c = 0$$

we obtain

$$ax - ax_0 + by - by_0 = 0; \quad y - y_0 = \frac{a(x_0 - x)}{b}.$$

Since $y - y_0$ is an integer and a and b are relatively prime, it is necessary that b divide $x_0 - x$; that is, $x_0 - x$ has the form

$$x_0 - x = bt,$$

where t is an integer. But then

$$y - y_0 = \frac{abt}{b} = at,$$

and we obtain

$$x = x_0 - bt, \quad y = y_0 + at.$$

Thus it is proved that every solution $[x,y]$ has the form *(4)*. It still remains to be verified that every pair of numbers $[x,y]$ obtained from *(4)* for integral $t = t_1$ will be a solution of *(3)*. In order to draw this inference, we substitute the quantities $x_1 = x_0 - bt_1$, $y_1 = y_0 + at_1$ into the left side of equation *(3)*:

---

* A pair of integers x and y which satisfy the equation we shall call a *solution* and write $[x, y]$.

$$ax_1 + by_1 + c = ax_0 - abt_1 + by_0 + abt_1 + c = ax_0 + by_0 + c.$$

But since $[x_0, y_0]$ is a solution, $ax_0 + by_0 + c = 0$ and, consequently,

$$ax_1 + by_1 + c = 0;$$

that is, $[x_1, y_1]$ is a solution of *(3)* and the theorem is completely proved.

Thus, if one solution of the equation $ax + by + c = 0$ is known, then all the remaining solutions are found from the arithmetic progression, the general term of which has the form

$$x = x_0 - bt, \quad y = y_0 + at \qquad (t = 0, \pm 1, \pm 2, \ldots).$$

We note that in the case $c = 0$, the solutions

$$x = -bt, \quad y = at$$

found earlier may be obtained from the formulas

$$x = x_0 - bt, \quad y = y_0 + at$$

if we choose $x_0 = y_0 = 0$. It is possible to make this choice since the pair $x = 0$, $y = 0$ is evidently a solution of the equation $ax + by = 0$.

How is a solution $[x_0, y_0]$ of equation *(3)* found in the general case when $c \neq 0$? Let us begin with an example—with the equation

$$127x - 52y + 1 = 0.$$

We shall transform the ratio of the coefficients of the unknowns.

First, we separate out the integral part of the improper fraction $\dfrac{127}{52}$:

$$\frac{127}{52} = 2 + \frac{23}{52}.$$

The proper fraction $\dfrac{23}{52}$ we replace by the equal fraction $\dfrac{1}{\dfrac{52}{23}}$.

Then we obtain

$$\frac{127}{52} = 2 + \frac{1}{\dfrac{52}{23}}.$$

We perform the same transformation on the improper fraction $\dfrac{52}{23}$ obtained in the denominator:

$$\frac{52}{23} = 2 + \frac{6}{23} = 2 + \frac{1}{\dfrac{23}{6}}.$$

Now the original fraction takes the form

$$\frac{127}{52} = 2 + \frac{1}{2 + \dfrac{1}{\dfrac{23}{6}}}.$$

We repeat the same argument for the fraction $\dfrac{23}{6}$:

$$\frac{23}{6} = 3 + \frac{5}{6} = 3 + \frac{1}{\dfrac{6}{5}}.$$

Then

$$\frac{127}{52} = 2 + \frac{1}{2 + \dfrac{1}{3 + \dfrac{1}{\dfrac{6}{5}}}}.$$

Separating out the integral part of the fraction $\dfrac{6}{5}$,

$$\frac{6}{5} = 1 + \frac{1}{5},$$

leads to the final result

$$\frac{127}{52} = 2 + \cfrac{1}{2 + \cfrac{1}{3 + \cfrac{1}{1 + \cfrac{1}{5}}}}$$

We have obtained an expression which is called a *finite continued fraction*. We remove the last term, $\frac{1}{5}$, of this continued fraction, convert the resulting continued fraction into a simple fraction, and subtract from the original fraction $\frac{127}{52}$:

$$2 + \cfrac{1}{2 + \cfrac{1}{3 + \cfrac{1}{1}}} = 2 + \cfrac{1}{2 + \cfrac{1}{4}} = 2 + \frac{4}{9} = \frac{22}{9},$$

$$\frac{127}{52} - \frac{22}{9} = \frac{1143 - 1144}{52 \cdot 9} = -\frac{1}{52 \cdot 9}.$$

If we bring the resulting equation to a common denominator and then reject the denominator, we obtain

$$127 \cdot 9 - 52 \cdot 22 + 1 = 0.$$

From a comparison of this equality with the equation

$$127x - 52y + 1 = 0$$

it follows that $x = 9, y = 22$ is a solution of the equation, and according to the theorem all of its solutions will be contained in the progression

$$x = 9 + 52t, \quad y = 22 + 127t \qquad (t = 0, \pm 1, \pm 2, \ldots).$$

The result leads to the thought that in the general case in order to find a solution of the equation $ax + by + c = 0$ it is sufficient to decompose the quotient of the coefficients of the unknowns into a continued fraction, reject its last term, and perform a calculation similar to that which was carried out above.

For a proof of this supposition, we shall need certain properties of continued fractions.

We consider the irreducible fraction $\dfrac{a}{b}$. We designate the integral part by q_1 and the remainder by r_2 when we divide a by b. We obtain

$$a = q_1 b + r_2, \qquad r_2 < b.$$

Let, further, q_2 and r_3 be the integral part and remainder respectively when we divide b by r_2. Then

$$b = q_2 r_2 + r_3, \qquad r_3 < r_2,$$

and in the same way

$$r_2 = q_3 r_3 + r_4, \qquad r_4 < r_3$$
$$r_3 = q_4 r_4 + r_5, \qquad r_5 < r_4$$
$$\cdot \ \cdot \ \cdot \ \cdot \ \cdot \ \cdot \ \cdot \ \cdot \ \cdot \ \cdot \ \cdot$$

The quantities q_1, q_2, ..., are called the *partial quotients*. The above process for the formation of the partial quotients is called the *Euclidean algorithm*. The remainders $r_2, r_3, \ldots,$ satisfy the inequalities

$$b > r_2 > r_3 > r_4 > \ldots \geq 0, \tag{5}$$

that is, they form a decreasing sequence of non-negative integers. Since the number of non-negative integers not exceeding b is finite, the above process must stop at a certain place because of the vanishing of the remainder r. Let r_n be the last non-zero remainder in the sequence (5). Then $r_{n+1} = 0$ and the Euclidean algorithm for the numbers a and b takes the form

$$a = q_1 b + r_2,$$
$$b = q_2 r_2 + r_3,$$
$$r_2 = q_3 r_3 + r_4,$$
$$\cdot \quad \cdot \quad \cdot \quad \cdot \quad \cdot \quad \cdot \quad \cdot \qquad (6)$$
$$r_{n-2} = q_{n-1} r_{n-1} + r_n,$$
$$r_{n-1} = q_n r_n.$$

We transcribe these equations into the form

$$\frac{a}{b} = q_1 + \frac{1}{\dfrac{b}{r_2}},$$

$$\frac{b}{r_2} = q_2 + \frac{1}{\dfrac{r_2}{r_3}},$$

$$\cdot \quad \cdot \quad \cdot \quad \cdot \quad \cdot \quad \cdot$$

$$\frac{r_{n-2}}{r_{n-1}} = q_{n-1} + \frac{1}{\dfrac{r_{n-1}}{r_n}},$$

$$\frac{r_{n-1}}{r_n} = q_n.$$

Replacing the value $\dfrac{b}{r_2}$ in the first line by its equal from the second line, the value $\dfrac{r_2}{r_3}$ by its equal from the third line, and so forth, we obtain an expansion of $\dfrac{a}{b}$ in a continued fraction

$$\frac{a}{b} = q_1 + \cfrac{1}{q_2 + \cfrac{1}{q_3 + \cfrac{}{\ddots + \cfrac{1}{q_{n-1} + \cfrac{1}{q_n}}}}}.$$

The expression obtained from a continued fraction by rejecting all of its terms beginning with a certain term we call a *con-*

vergent. We obtain the first convergent δ_1 by rejecting all terms beginning with $\dfrac{1}{q_2}$:

$$\delta_1 = q_1 < \frac{a}{b} .$$

The second convergent δ_2 is obtained by rejecting all terms beginning with $\dfrac{1}{q_3}$:

$$\delta_2 = q_1 + \frac{1}{q_2} > \frac{a}{b} .$$

In the same way,

$$\delta_3 = q_1 + \cfrac{1}{q_2 + \cfrac{1}{q_3}} < \frac{a}{b} ,$$

$$\delta_4 = q_1 + \cfrac{1}{q_2 + \cfrac{1}{q_3 + \cfrac{1}{q_4}}} > \frac{a}{b} ,$$

and so forth.

As a consequence of the method of formation of the convergents, we evidently have the inequalities

$$\delta_1 < \delta_3 \ldots < \delta_{2k-1} < \frac{a}{b} ; \quad \delta_2 > \delta_4 > \ldots > \delta_{2k} > \frac{a}{b} .$$

We write the kth convergent δ_k in the form

$$\delta_k = \frac{P_k}{Q_k} \qquad (1 \leq k \leq n)$$

and seek the law of formation of the numerator and denominator of this convergent. We transform first the convergents δ_1, δ_2, and δ_3:

$$\delta_1 = q_1 = \frac{q_1}{1} = \frac{P_1}{Q_1}; \qquad\qquad P_1 = q_1; \qquad\qquad Q_1 = 1;$$

$$\delta_2 = q_1 + \frac{1}{q_2} = \frac{q_1 q_2 + 1}{q_2} = \frac{P_2}{Q_2}; \quad P_2 = q_1 q_2 + 1; \quad Q_2 = q_2;$$

$$\delta_3 = q_1 + \cfrac{1}{q_2 + \cfrac{1}{q_3}} = q_1 + \frac{q_3}{q_2 q_3 + 1} = \frac{q_1 q_2 q_3 + q_1 + q_3}{q_2 q_3 + 1} = \frac{P_3}{Q_3};$$

$$P_3 = q_1 q_2 q_3 + q_1 + q_3; \quad Q_3 = q_2 q_3 + 1.$$

From this we obtain

$$P_3 = P_2 q_3 + P_1; \quad Q_3 = Q_2 q_3 + Q_1.$$

We shall show by induction * that the corresponding forms

$$P_k = P_{k-1} q_k + P_{k-2}, \quad Q_k = Q_{k-1} q_k + Q_{k-2} \qquad (7)$$

are satisfied for all $k \geq 3$.

Indeed, let the equalities *(7)* be satisfied for a certain $k \geq 3$. From the definition of a convergent, it follows immediately that if we replace in δ_k the number q_k by $q_k + \frac{1}{q_{k+1}}$, then δ_k is transformed into δ_{k+1}. According to the induction hypothesis,

$$\delta_k = \frac{P_k}{Q_k} = \frac{P_{k-1} q_k + P_{k-2}}{Q_{k-1} q_k + Q_{k-2}}.$$

Replacing q_k by $q_k + \frac{1}{q_{k+1}}$ we obtain

$$\delta_{k+1} = \frac{P_{k-1}\left(q_k + \cfrac{1}{q_{k+1}}\right) + P_{k-2}}{Q_{k-1}\left(q_k + \cfrac{1}{q_{k+1}}\right) + Q_{k-2}} = \frac{P_k + \cfrac{1}{q_{k+1}} P_{k-1}}{Q_k + \cfrac{1}{q_{k+1}} Q_{k-1}}$$

$$= \frac{P_k q_{k+1} + P_{k-1}}{Q_k q_{k+1} + Q_{k-1}}.$$

---

* See the book in this series by I. S. Sominski, *Method of Mathematical Induction*, 1950. (This book is in Russian, but the reader can consult almost any college algebra text for details on mathematical induction [*Translator*].)

Hence, since

$$\delta_{k+1} = \frac{P_{k+1}}{Q_{k+1}},$$

it follows that

$$P_{k+1} = P_k q_{k+1} + P_{k-1}, \ Q_{k+1} = Q_k q_{k+1} + Q_{k-1}.$$

Thus from the truth of equations (7) for a certain $k \geq 3$, the truth for $k + 1$ follows. But for $k = 3$ the equations (7) are true, and consequently the equalities are established for all $k \geq 3$.

We shall now show that the difference of neighboring convergents $\delta_k - \delta_{k-1}$ satisfies the relation

$$\delta_k - \delta_{k-1} = \frac{(-1)^k}{Q_k Q_{k-1}} \qquad (k > 1). \tag{8}$$

Indeed,

$$\delta_k - \delta_{k-1} = \frac{P_k}{Q_k} - \frac{P_{k-1}}{Q_{k-1}} = \frac{P_k Q_{k-1} - Q_k P_{k-1}}{Q_k Q_{k-1}}.$$

Using equations (7) we transform the numerator of this new fraction:

$$P_k Q_{k-1} - Q_k P_{k-1} = (P_{k-1} q_k + P_{k-2}) Q_{k-1} - (Q_{k-1} q_k + Q_{k-2}) P_{k-1}$$
$$= -(P_{k-1} Q_{k-2} - Q_{k-1} P_{k-2}).$$

The expression in the parentheses may be obtained from the original one by replacing k by $k - 1$. Repeating this transformation for the new expression, we evidently obtain the chain

$$P_k Q_{k-1} - Q_k P_{k-1} = (-1)(P_{k-1} Q_{k-2} - Q_{k-1} P_{k-2})$$
$$= (-1)^2 (P_{k-2} Q_{k-3} - Q_{k-2} P_{k-3})$$
$$= \ldots = (-1)^{k-2} (P_2 Q_1 - Q_2 P_1)$$
$$= (-1)^{k-2} (q_1 q_2 + 1 - q_2 q_1) = (-1)^{k-2}.$$

Whence it follows that

$$\delta_k - \delta_{k-1} = \frac{P_k Q_{k-1} - Q_k P_{k-1}}{Q_k Q_{k-1}} = \frac{(-1)^{k-2}}{Q_k Q_{k-1}} = \frac{(-1)^k}{Q_k Q_{k-1}}.$$

If the expansion of $\frac{a}{b}$ into a continued fraction has n terms, then the nth convergent δ_n coincides with $\frac{a}{b}$. Applying (8) when $k = n$ yields

$$\delta_n - \delta_{n-1} = \frac{(-1)^n}{Q_n Q_{n-1}},$$

$$\frac{a}{b} - \delta_{n-1} = \frac{(-1)^n}{b Q_{n-1}}.$$

$$(9)$$

We return to the solution of the equation

$$ax + by + c = 0, \qquad (a, b) = 1. \qquad (10)$$

We rewrite the relation (9) in the form

$$\frac{a}{b} - \frac{P_{n-1}}{Q_{n-1}} = \frac{(-1)^n}{b Q_{n-1}}.$$

Reducing this to a common denominator and rejecting it, we obtain

$$aQ_{n-1} - bP_{n-1} = (-1)^n,$$
$$aQ_{n-1} + b(-P_{n-1}) + (-1)^{n-1} = 0.$$

We multiply this relation by $(-1)^{n-1}c$. Then

$$a[(-1)^{n-1}c Q_{n-1}] + b[(-1)^n c P_{n-1}] + c = 0.$$

Hence it follows that *the pair of integers* $[x_0, y_0]$,

$$x_0 = (-1)^{n-1}c Q_{n-1}, \quad y_0 = (-1)^n c P_{n-1}, \qquad (11)$$

is a solution of the equation (10), and, according to the theorem, all solutions of this equation have the form

$$x = (-1)^{n-1}cQ_{n-1} - bt, \quad y = (-1)^n cP_{n-1} + at$$
$$(t = 0, \pm 1, \pm 2, \dots).$$

The results obtained completely solve the problem of finding all integral solutions of a first degree equation in two unknowns. We proceed now to consider certain equations of second degree.

3. Examples of Equations of Second Degree in Three Unknowns

EXAMPLE A. We consider the equation

$$x^2 + y^2 = z^2 \tag{12}$$

of second degree in three unknowns. Geometrically we may interpret the solving of this equation in integers as the finding of all Pythagorean triangles, that is, right triangles whose sides $x, y,$ and z are integers.

We designate the greatest common divisor of x and y by $d = (x, y)$. Then

$$x = x_1 d, \quad y = y_1 d$$

and equation (12) takes the form

$$x_1^2 d^2 + y_1^2 d^2 = z^2.$$

Hence it follows that d^2 divides z^2 and this means z is a multiple of d: $z = z_1 d$.

We now may write (12) in the form

$$x_1^2 d^2 + y_1^2 d^2 = z_1^2 d^2.$$

Canceling d^2 we obtain

$$x_1^2 + y_1^2 = z_1^2.$$

We have arrived at an equation of the same form as the original, where now the integers x_1 and y_1 do not have any common divisor other than 1. Thus in solving *(12)* we may restrict our considerations to the case where x and y are relatively prime. Thus we let $(x, y) = 1$. Then either x or y (say x) will be odd. Transferring y^2 to the right side of *(12)*, we obtain:

$$x^2 = z^2 - y^2; \quad x^2 = (z + y)(z - y). \tag{13}$$

We designate the greatest common divisor of the expressions $z + y, z - y$ by d_1. Then

$$z + y = ad_1, \quad z - y = bd_1, \tag{14}$$

where a and b are relatively prime.

Substituting into *(13)* these values of $z + y, z - y$ yields

$$x^2 = abd_1^2.$$

Since the numbers a and b have no common divisors, the equality obtained is possible only where a and b are perfect squares: *

$$a = u^2, \quad b = v^2.$$

But then

$$x^2 = u^2 v^2 d_1^2$$

and

$$x = uvd_1. \tag{15}$$

We find y and z now from equality *(14)*. Adding these equations gives

$$2z = ad_1 + bd_1 = u^2 d_1 + v^2 d_1; \quad z = \frac{u^2 + v^2}{2} d_1. \tag{16}$$

Subtracting the second of equalities *(14)* from the first yields

$$2y = ad_1 - bd_1 = u^2 d_1 - v^2 d_1; \quad y = \frac{u^2 - v^2}{2} d_1. \tag{17}$$

---

*It is known that the product of two relatively prime numbers may be a perfect square only when each factor is a perfect square.

Since x is odd we obtain from *(15)* that u, v, and d_1 are also odd. Moreover $d_1 = 1$, otherwise it would follow from the equalities

$$x = uvd_1 \quad \text{and} \quad y = \frac{u^2 - v^2}{2}d_1$$

that the quantities x and y have a common divisor $d_1 \neq 1$, which is contrary to the supposition that they are relatively prime. The numbers u and v are connected with the relatively prime numbers a and b by the equalities

$$a = u^2, \quad b = v^2,$$

and in consequence of this are themselves relatively prime; $v < u$ since $b < a$, which is clear from equality *(14)*.

Substituting $d_1 = 1$ into equations *(15)*, *(16)*, and *(17)*, we obtain the *formulas*

$$x = uv, \quad y = \frac{u^2 - v^2}{2}, \quad z = \frac{u^2 + v^2}{2}, \tag{18}$$

giving for odd relatively prime u and v (v < u) all triples of positive integers x, y, z which have no common divisor other than unity and which satisfy equation (12). By simple substitution of x, y, and z into *(12)*, it is easy to verify that for any numbers u and v formulas *(18)* satisfy this equation.

For the first values of u and v, formulas *(18)* lead to the following frequently met equations:

$$3^2 + 4^2 = 5^2 \qquad (v = 1, \quad u = 3),$$
$$5^2 + 12^2 = 13^2 \qquad (v = 1, \quad u = 5),$$
$$15^2 + 8^2 = 17^2 \qquad (v = 3, \quad u = 5).$$

As we have already noted, formulas *(18)* give only those solutions of the equation

$$z^2 + y^2 = z^2$$

in which the numbers x, y, and z do not have common divisors other than unity. All other positive integral solutions of this equation are obtained by multiplying the solutions obtained in formulas *(18)* by a common multiplier d.

In the same way that we have obtained all solutions of *(12)*, we can obtain all solutions of other equations of the same type.

EXAMPLE B. We shall find all solutions of the equation

$$x^2 + 2y^2 = z^2 \tag{19}$$

in positive integers x, y, z which are relatively prime in pairs.

We note that if $[x, y, z]$ is a solution of *(19)* and x, y, z do not have a common divisor other than 1, then they are relatively prime in pairs. Indeed, if x and y are multiples of a prime number p, then from the equality

$$\left(\frac{x}{p}\right)^2 + 2\left(\frac{y}{p}\right)^2 = \left(\frac{z}{p}\right)^2$$

it follows, since the left side is an integer, that z is a multiple of p. The same will be true if x and z or y and z are divisible by p.

It is necessary, we note, for x to be odd in order that the greatest common divisor of x, y, z will equal 1. Indeed, if x is even, then the left side of *(19)* will be even, and this means that z will also be even. But x^2 and z^2 will then be multiples of 4. Hence it follows that $2y^2$ must be divisible by 4; in other words, that y must also be even. This means, if x is even, then all of the numbers x, y, z must be even. Thus in a solution with no common divisors other than 1, x must be odd. From this it follows that z must also be odd. Transferring x^2 to the right side, we obtain

$$2y^2 = z^2 - x^2 = (z + x)(z - x).$$

But $z + x$ and $z - x$ have greatest common divisor 2. Indeed, let their greatest common divisor be d.* Then

$$z + x = kd, \quad z - x = ld,$$

---

*$d \geqq 2$ since $z + x$ and $z - x$ are even [*Tr.*].

where k and l are integers. Adding and subtracting these equations, we have

$$2z = d(k + l), \quad 2x = d(k - l).$$

But z and x are odd and relatively prime. Therefore the greatest common divisor of $2x$ and $2z$ will be 2.[*] Hence it follows that $d = 2$.

Thus either

$$\frac{z + x}{2} \quad \text{or} \quad \frac{z - x}{2}$$

is odd. Therefore either

$$z + x \quad \text{and} \quad \frac{z - x}{2}$$

are relatively prime or

$$\frac{z + x}{2} \quad \text{and} \quad z - x$$

are relatively prime. In the first case from the equality

$$(z + x) \cdot \frac{z - x}{2} = y^2$$

it follows that

$$z + x = n^2, \quad z - x = 2m^2,$$

and in the second case from the equality

$$\frac{z + x}{2}(z - x) = y^2$$

it follows that

$$z + x = 2m^2, \quad z - x = n^2,$$

where n and m are integers, m is odd, and $n > 0, m > 0$. Solving these two systems of equations for x and z and finding y, we obtain either

---

[*] This with $2z = d(k + l)$, $2x = d(k - l)$ shows $d \leqq 2$ [*Tr.*].

$$z = \frac{1}{2}(n^2 + 2m^2), \quad x = \frac{1}{2}(n^2 - 2m^2), \quad y = mn$$

or

$$z = \frac{1}{2}(n^2 + 2m^2), \quad x = \frac{1}{2}(2m^2 - n^2), \quad y = mn$$

where m is odd. Unifying these two formulas representing solutions x, y, z we obtain the general formulas

$$x = \pm\frac{1}{2}(n^2 - 2m^2), \quad y = mn, \quad z = \frac{1}{2}(n^2 + 2m^2)$$

where m is odd. But in order that z and x will be integers it is necessary that n be even. Assuming $n = 2b$ and $m = a$, we obtain the final *general formulas giving all solutions of equation (19) in positive integers x, y, z having no common factor larger than 1:*

$$x = \pm(a^2 - 2b^2), \quad y = 2ab, \quad z = a^2 + 2b^2, \qquad (19')$$

where a and b are positive, odd, relatively prime integers. Here a and b are chosen in any way so that x will be positive. Formulas (19') indeed give all solutions in positive relatively prime integers x, y, z since, on the one hand, we show that x, y, z in this case must be be represented by formulas (19'). And, on the other hand, if we are given numbers a and b satisfying our conditions, then x, y, z will indeed be relatively prime and will be solutions of equations (19).

4. Finding All Solutions of Equations of the Form $x^2 - Ay^2 = 1$

Let us now consider the solving in integers of those equations of the second degree in two unknowns having the form

$$x^2 - Ay^2 = 1, \qquad (20)$$

where A is a positive integer which is not a perfect square. To solve these equations, we must first acquaint ourselves with the continued fraction expansions of irrational numbers of the form $\sqrt{A}$. From the Euclidean algorithm it follows that each rational number expands into a continued fraction with a finite number of terms. An irrational number, however, expands into a continued fraction with an infinite number of terms. For example, let us find the continued fraction expansion of $\sqrt{2}$.

We transform the evident identities

$$(\sqrt{2} - 1)(\sqrt{2} + 1) = 1,$$

$$\sqrt{2} - 1 = \frac{1}{\sqrt{2} + 1},$$

$$\sqrt{2} - 1 = \frac{1}{2 + (\sqrt{2} - 1)};$$

replacing the difference $\sqrt{2} - 1$ obtained in the denominator by the equal expression

$$\frac{1}{2 + (\sqrt{2} - 1)},$$

we obtain

$$\sqrt{2} - 1 = \frac{1}{2 + \dfrac{1}{2 + (\sqrt{2} - 1)}},$$

$$\sqrt{2} = 1 + \frac{1}{2 + \dfrac{1}{2 + (\sqrt{2} - 1)}}.$$

Again we replace the bracketed expression in the denominator of the last equality by its equal above and get

$$\sqrt{2} = 1 + \frac{1}{2 + \dfrac{1}{2 + \dfrac{1}{2 + (\sqrt{2} - 1)}}}$$

Continuing this process, we obtain the following expansion of $\sqrt{2}$ in an infinite continued fraction:

$$\sqrt{2} = 1 + \cfrac{1}{2 + \cfrac{1}{2 + \cfrac{1}{2 + \cfrac{1}{2 + \cfrac{1}{2 + \cdots}}}}} \qquad (21)$$

We notice that the above method uses an identity of the form

$$(\sqrt{m^2 + 1} - m)(\sqrt{m^2 + 1} + m) = 1,$$

which is not suitable for every irrational $\sqrt{A}$. This method may evidently be used when the integer A can be put in the form $A = m^2 + 1$, where m is a non-zero integer. (In particular, if $m = 1$, we obtain the expansion for $\sqrt{2}$; if $m = 2$, the expansion for $\sqrt{5}$; and so forth.) However, in the general case, a comparatively simple method for expanding $\sqrt{A}$ into an infinite continued fraction is also known.*

We form for the infinite continued fraction *(21)*, as we did for the finite ones, the sequence of convergents $\delta_1, \delta_2, \delta_3, \ldots$

$$\delta_1 = 1, \qquad\qquad\qquad \delta_1 < \sqrt{2}\,;$$

$$\delta_2 = 1 + \frac{1}{2} = \frac{3}{2}, \qquad \delta_2 > \sqrt{2}\,;$$

$$\delta_3 = 1 + \cfrac{1}{2 + \cfrac{1}{2}} = \frac{7}{5}, \quad \delta_3 < \sqrt{2} \qquad (22)$$

$$\delta_4 = \ldots = \frac{17}{12}, \qquad\quad \delta_4 > \sqrt{2}\,;$$

and so forth.

---

* See, for example, the book *Theory of Numbers* by Y. V. Arnold, chapter VI, 1939, or *Continued Fractions* by A. Y. Khinchin, 1949. (These books are in Russian. The reader may consult, for example, *The Higher Arithmetic* by H. Davenport or *Introduction to the Theory of Numbers* by G. H. Hardy and E. M. Wright [*Tr.*]).

From the method of formation of the convergents it follows that

$$\delta_1 < \delta_3 < \ldots < \sqrt{2},$$
$$\delta_2 > \delta_4 > \ldots > \sqrt{2}.$$

In general, for a given expansion of an irrational number α into an infinite continued fraction

$$\alpha = q_1 + \cfrac{1}{q_2 + \cfrac{1}{q_3 + \cfrac{1}{\ddots}}}$$

the convergents satisfy the inequalities

$$\delta_1 < \delta_3 < \ldots < \delta_{2k+1} < \ldots < \alpha < \ldots < \delta_{2k} < \ldots < \delta_4 < \delta_2. \quad (23)$$

We represent the convergent δ_k by

$$\delta_k = \frac{P_k}{Q_k}.$$

The relations *(7)*

$$P_k = P_{k-1}q_k + P_{k-2}, \quad Q_k = Q_{k-1}q_k + Q_{k-2},$$

obtained previously for finite continued fractions, remain valid for infinite continued fractions, since in the derivation of these relations we nowhere exploited the fact that the continued fractions were finite. Therefore relation *(8)* between neighboring convergents is also preserved:

$$\delta_k - \delta_{k-1} = \frac{(-1)^k}{Q_k Q_{k-1}}. \quad (24)$$

For example, for the convergents of the expansion of $\sqrt{2}$ into a continued fraction, we obtain from *(22)* for $k = 3$ and $k = 4$

$$\delta_3 - \delta_2 = \frac{7}{5} - \frac{3}{2} = \frac{-1}{10},$$

$$\delta_4 - \delta_3 = \frac{17}{12} - \frac{7}{5} = \frac{1}{60}.$$

These, naturally, coincide with the results indicated by equation *(24)*.

From equation *(24)* in particular, it follows that

$$\delta_{2k} - \delta_{2k+1} = -\left(\delta_{2k+1} - \delta_{2k}\right) = -\frac{(-1)^{2k+1}}{Q_{2k+1}Q_{2k}} = \frac{1}{Q_{2k+1}Q_{2k}}.$$

We show now that the inequality,

$$0 < P_{2k} - \alpha Q_{2k} < \frac{1}{Q_{2k+1}}, \qquad (25)$$

is valid. Indeed, the left side of this inequality is easily obtained since, according to *(23)*,

$$\alpha < \delta_{2k} = \frac{P_{2k}}{Q_{2k}}; \quad \alpha Q_{2k} < P_{2k}; \quad 0 < P_{2k} - \alpha Q_{2k}.$$

The proof of the right part of inequality *(25)* also proceeds without difficulty. In virtue of *(23)*,

$$\delta_{2k+1} < \alpha < \delta_{2k};$$

consequently,

$$\delta_{2k} - \alpha < \delta_{2k} - \delta_{2k+1} = \frac{1}{Q_{2k}Q_{2k+1}}.$$

Hence, replacing δ_{2k} by $\dfrac{P_{2k}}{Q_{2k}}$, we obtain

$$\frac{P_{2k}}{Q_{2k}} - \alpha < \frac{1}{Q_{2k}Q_{2k+1}}.$$

Multiplying this inequality by Q_{2k} leads to the desired result,

$$P_{2k} - \alpha Q_{2k} < \frac{1}{Q_{2k+1}}.$$

We now apply the result to the solution of the equation

$$x^2 - 2y^2 = 1. \qquad (26)$$

Transforming the left side of this equation, we get

$$x^2 - 2y^2 = (x - \sqrt{2}y)(x + \sqrt{2}y).$$

We put $x = P_{2k}$ and $y = Q_{2k}$, where P_{2k} and Q_{2k} are the numerator and denominator of the corresponding convergent of the continued fraction expansion of $\sqrt{2}$. Then

$$P_{2k}^2 - 2Q_{2k}^2 = (P_{2k} - \sqrt{2}Q_{2k})(P_{2k} + \sqrt{2}Q_{2k}). \qquad (27)$$

On the left, and this means also on the right, side of the equality, we have an integer. We show that this integer is greater than zero but less than 2; consequently it is equal to one. For this we apply inequality *(25)* for $\alpha = \sqrt{2}$:

$$0 < P_{2k} - \sqrt{2}Q_{2k} < \frac{1}{Q_{2k+1}}. \qquad (28)$$

Thus, since both factors of the right side of *(27)* are positive, we have

$$P_{2k}^2 - 2Q_{2k}^2 > 0.$$

On the other hand,

$$P_{2k} - \sqrt{2}Q_{2k} < \frac{1}{Q_{2k+1}} = \frac{1}{Q_{2k}q_{2k+1} + Q_{2k-1}}$$

$$= \frac{1}{2Q_{2k} + Q_{2k-1}} < \frac{1}{2Q_{2k}}.$$

But by *(23)*,

$$\delta_{2k} = \frac{P_{2k}}{Q_{2k}} > \sqrt{2}.$$

Hence

$$\sqrt{2}Q_{2k} < P_{2k},$$

$$P_{2k} + \sqrt{2}Q_{2k} < 2P_{2k},$$

and we obtain two inequalities for the factors on the right side of *(27)*:

$$P_{2k} - \sqrt{2}\,Q_{2k} < \frac{1}{2Q_{2k}},$$

$$P_{2k} + \sqrt{2}\,Q_{2k} < 2P_{2k}.$$

Multiplying these inequalities together gives

$$P_{2k}^2 - 2Q_{2k}^2 < \frac{P_{2k}}{Q_{2k}}.$$

Applying inequality *(28)*, we obtain

$$P_{2k}^2 - 2Q_{2k}^2 < \frac{\sqrt{2}\,Q_{2k} + \dfrac{1}{Q_{2k+1}}}{Q_{2k}} = \sqrt{2} + \frac{1}{Q_{2k}Q_{2k+1}},$$

and since for $k \geq 1$

$$\frac{1}{Q_{2k}Q_{2k+1}} \leq \frac{1}{Q_2 Q_3} = \frac{1}{10}$$

then

$$P_{2k}^2 - 2Q_{2k}^2 < \sqrt{2} + \frac{1}{10} < 2.$$

Thus we have shown that the integer $P_{2k}^2 - 2Q_{2k}^2$, for all $k \geq 1$, satisfies the inequality

$$0 < P_{2k}^2 - 2Q_{2k}^2 < 2.$$

Consequently

$$P_{2k}^2 - 2Q_{2k}^2 = 1,$$

that is, the numbers $x = P_{2k}, y = Q_{2k}$ for arbitrary $k \geq 1$ give a solution of the equation

$$x^2 - 2y^2 = 1.$$

For the time being we do not know whether the solutions found for the equation *(26)* represent all the solutions of this equation.

It is natural now to raise the question of how to obtain all integral solutions x, y of the equation

$$x^2 - Ay^2 = 1, \qquad (29)$$

where $A > 0$ is integral and $\sqrt{A}$ is irrational. * As we shall show, all solutions can be obtained if one solution of equation *(29)* is known. For example, equation *(26)*, we have seen, has a solution. We concern ourselves now with the question of how to obtain all solutions of equation *(29)* from one definite solution, which we shall call the smallest or minimal one; for the present, the question remains whether equation *(29)* always has a solution in integers different from the trivial solution $x = 1, y = 0$.

Let us assume that equation *(29)* has a non-trivial solution $[x_0, y_0]$, $x_0 > 0$, $y_0 > 0$ and

$$x_0^2 - Ay_0^2 = 1. \qquad (30)$$

(We remember that a solution is a pair of integers $[x_0, y_0]$ satisfying the equation.) We shall call that solution $[x_0, y_0]$ *minimal* if for $x = x_0$ and $y = y_0$ the binomial $x + \sqrt{A}y$, $\sqrt{A} > 0$ has the smallest possible size of all possible values which it assumes for x and y running over all positive integer solutions of equation *(29)*. For example, for equation *(26)* the minimal solution will be $x = 3$, $y = 2$ since $x + \sqrt{2}y$ for these values of x and y takes the value $3 + 2\sqrt{2}$, and there does not exist another solution of *(26)* which does not yield a larger value for $x + \sqrt{2}y$. This is seen at once if one attempts to choose a smaller pair of integers which give a solution. Indeed, the next, in magnitude, solution of equation *(26)* is $x = 17, y = 12$, and clearly $17 + 12\sqrt{2}$ is larger than $3 + 2\sqrt{2}$. We

---

* Equation *(29)* is often called Pell's equation [*Tr.*].

notice, also, that there do not exist two minimal solutions of equation *(29)*. Let us assume the contrary, that is, that there are solutions $[x_1, y_1]$ and $[x_2, y_2]$ which give one and the same value to the binomial $x + \sqrt{A}y$. Then

$$x_1 + \sqrt{A}y_1 = x_2 + \sqrt{A}y_2. \tag{31}$$

But $\sqrt{A}$ is an irrational number and x_1, y_1, x_2, y_2 are integers. From *(31)* it follows immediately that

$$x_1 - x_2 = (y_2 - y_1)\sqrt{A}.$$

This is not possible since $x_1 - x_2$ is an integer and $(y_2 - y_1)\sqrt{A}$, being a product of an integer and an irrational number, is an irrational number, and no integer is irrational. This contradiction vanishes if $x_1 = x_2$ and $y_1 = y_2$; in other words, when we had not two different solutions but one. Thus, if there is a minimal solution, there is only one. We now note one more very important property of solutions of equation *(29)*. Let $[x_1, y_1]$ be a solution of *(29)*. Then

$$x_1^2 - Ay_1^2 = 1$$

or

$$(x_1 + \sqrt{A}y_1)(x_1 - \sqrt{A}y_1) = 1. \tag{32}$$

Let us now take the two sides of the equality *(32)* to the positive integral power n:

$$(x_1 + \sqrt{A}y_1)^n(x_1 - \sqrt{A}y_1)^n = 1. \tag{33}$$

Expanding this by the binomial theorem, we have

$$(x_1 + \sqrt{A}y_1)^n = x_1^n + nx_1^{n-1}\sqrt{A}y_1 + \frac{n(n-1)}{2}x_1^{n-2}Ay_1^2$$
$$+ \ldots + (\sqrt{A})^n y_1^n = x_n + \sqrt{A}y_n, \tag{34}$$

where x_n and y_n are integers since the first, third, and generally all the odd terms in this binomial expansion are integers and the even terms are integers multiplied by $\sqrt{A}$. Collecting into separate parts the summands of integers and of integral multiples of $\sqrt{A}$ we obtain equation *(34)*. The numbers x_n and y_n, as we show next, are also solutions of equation *(29)*. Indeed, from equation *(34)*, changing the sign of $\sqrt{A}$, we obtain

$$(x_1 - \sqrt{A}y_1)^n = x_n - \sqrt{A}y_n \tag{35}$$

Multiplying equalities *(34)* and *(35)* and making use of equation *(33)*, we obtain finally

$$(x_1 + \sqrt{A}y_1)^n (x_1 - \sqrt{A}y_1)^n = (x_n + \sqrt{A}y_n)(x_n - \sqrt{A}y_n)$$
$$= x_n^2 - Ay_n^2 = 1; \tag{36}$$

in other words, $[x_n, y_n]$ *is also a solution of equation (29)*.

We may now prove the complete theorem relating to the solutions of equation *(29)*.

THEOREM II. *Every solution of equation (29)*

$$x^2 - Ay^2 = 1$$

for positive A and $\sqrt{A}$ irrational has the form $[\pm x_n, \pm y_n]$, *where*

$$x_n = \frac{1}{2}[(x_0 + y_0\sqrt{A})^n + (x_0 - y_0\sqrt{A})^n],$$
$$y_n = \frac{1}{2\sqrt{A}}[(x_0 + y_0\sqrt{A})^n - (x_0 - y_0\sqrt{A})^n], \tag{37}$$

where $[x_0, y_0]$ *is the minimal solution.*

Proof: Let us assume the converse, namely, that there exists a positive integral solution $[x', y']$ of equation *(29)* such that the equality

$$x' + \sqrt{A}y' = (x_0 + \sqrt{A}y_0)^n \tag{38}$$

is impossible for any positive integer n. We consider the series of numbers

$$x_0 + \sqrt{A}y_0, \ (x_0 + \sqrt{A}y_0)^2, \ (x_0 + \sqrt{A}y_0)^3, \ldots.$$

This is an increasing unbounded sequence of positive numbers since $x_0 \geq 1, y_0 \geq 1$, and $x_0 + \sqrt{A}y_0 > 1$. Since $[x_0, y_0]$ is the minimal solution, we have

$$x' + \sqrt{A}y' > x_0 + \sqrt{A}y_0.$$

Therefore there is always an integer $n \geq 1$ such that

$$(x_0 + \sqrt{A}y_0)^n < x' + \sqrt{A}y' < (x_0 + \sqrt{A}y_0)^{n+1}. \quad (39)$$

But $x_0 - \sqrt{A}y_0 > 0$ since

$$(x_0 + \sqrt{A}y_0)(x_0 - \sqrt{A}y_0) = x_0^2 - Ay_0^2 = 1 > 0.$$

Therefore multiplying all terms of inequality *(39)* by the positive number $(x_0 - \sqrt{A}y_0)^n$ preserves the inequality, and we obtain

$$(x_0 + \sqrt{A}y_0)^n(x_0 - \sqrt{A}y_0)^n < (x' + \sqrt{A}y')(x_0 - \sqrt{A}y_0)^n$$
$$< (x_0 + \sqrt{A}y_0)^{n+1}(x_0 - \sqrt{A}y_0)^n. \quad (40)$$

Since

$$(x_0 + \sqrt{A}y_0)^n(x_0 - \sqrt{A}y_0)^n = (x_0^2 - Ay_0^2)^n = 1, \quad (41)$$

then

$$(x_0 + \sqrt{A}y_0)^{n+1}(x_0 - \sqrt{A}y_0)^n = x_0 + \sqrt{A}y_0. \quad (42)$$

In addition,

$$(x' + \sqrt{A}y')(x_0 - \sqrt{A}y_0)^n = (x' + \sqrt{A}y')(x_n - \sqrt{A}y_n)$$
$$= x'x_n - Ay'y_n + \sqrt{A}(y'x_n - x'y_n) = \bar{x} + \sqrt{A}\bar{y}, \quad (43)$$

where $\bar{x}$ and $\bar{y}$ are integers and

$$x_n - \sqrt{A}y_n = (x_0 - \sqrt{A}y_0)^n.$$

Making use of relations *(41)*, *(42)*, *(43)*, and inequality *(40)*, we obtain the inequality

$$1 < \bar{x} + \sqrt{A}\bar{y} < x_0 + \sqrt{A}y_0. \qquad (44)$$

We shall show that the pair of integers $\bar{x}$ and $\bar{y}$ is a solution of equation *(29)*. Indeed, we multiply together the equality *(43)*, that is,

$$\bar{x} + \sqrt{A}\bar{y} = (x' + \sqrt{A}y')(x_0 - \sqrt{A}y_0)^n \qquad (45)$$

and the equality

$$\bar{x} - \sqrt{A}\bar{y} = (x' - \sqrt{A}y')(x_0 + \sqrt{A}y_0)^n, \qquad (46)$$

which is obtained from equation *(43)* if we change the sign of $y\sqrt{A}$, and we obtain

$$
\begin{aligned}
(\bar{x} + \sqrt{A}\bar{y})(\bar{x} - \sqrt{A}\bar{y}) &= \bar{x}^2 - A\bar{y}^2 \\
&= (x' + \sqrt{A}y')(x' - \sqrt{A}y')(x_0 + \sqrt{A}y_0)^n(x_0 - \sqrt{A}y_0)^n \quad (47) \\
&= (x'^2 - Ay'^2)(x_0^2 - Ay_0^2)^n = 1
\end{aligned}
$$

since $[x', y']$ and $[x_0, y_0]$ are solutions of *(29)*. We shall prove, finally, that $\bar{x} > 0$ and $\bar{y} > 0$. First of all, it is clear that $\bar{x}$ is not equal to zero. Indeed, if $\bar{x} = 0$ then from equation *(47)* we would have

$$-A\bar{y}^2 = 1,$$

which is a contradiction since $A > 0$. Further, if $\bar{y} = 0$ then $\bar{x}^2 = 1$, but from inequality *(44)* we have $\bar{x} > 1$, which is not possible. Finally, we note, that the signs of $\bar{x}$ and $\bar{y}$ must be the same. Indeed, if it is supposed that the signs of $\bar{x}$ and $\bar{y}$ are different, then $\bar{x}$ and $-\bar{y}$ will have the same sign. If we compare, then, the absolute values of the numbers $\bar{x} + \sqrt{A}\bar{y}$ and

$\bar{x} - \sqrt{A}\bar{y}$, we see that the absolute value of the first is the difference of the two numbers $\bar{x}$, $-\sqrt{A}\bar{y}$ (which have the same sign) whereas the second is the sum of these two numbers. But we also know that

$$\bar{x} + \sqrt{A}\bar{y} > 1;$$

this means $\bar{x} - \sqrt{A}\bar{y}$ is also of absolute value greater than unity.
But

$$(\bar{x} + \sqrt{A}\bar{y})(\bar{x} - \sqrt{A}\bar{y}) = \bar{x}^2 - A\bar{y}^2 = 1,$$

and we have reached a contradiction, since the product of two numbers each of which is of absolute value larger than unity must also have absolute value larger than unity. And so the signs of $\bar{x}$ and $\bar{y}$ are the same and $\bar{x} \neq 0$, $\bar{y} \neq 0$. But then from inequality *(44)*, it follows at once that $\bar{x} > 0$ and $\bar{y} > 0$. Thus, from the supposition that

$$x^2 - Ay^2 = 1, \qquad A > 0,$$

has a solution, $[x'\ y']$, not of the form given in *(38)*, we have found a positive solution $[\bar{x}, \bar{y}]$ satisfying inequality *(44)*, and this contradicts the minimality of $[x_0, y_0]$. With this we have proved that the supposed existence of a solution not representable by equation *(38)* leads us to a contradiction. In other words, we have shown that all solutions of our equation can be obtained from equation *(38)*.

Thus each solution $[x, y]$ of equation *(29)* is obtained from the relation

$$x + \sqrt{A}y = (x_0 + \sqrt{A}y_0)^n, \qquad n \geq 0, \qquad (48)$$

where $[x_0, y_0]$ is the minimal solution. Changing the sign of $y\sqrt{A}$ in this last equality yields the equality

$$x - \sqrt{A}y = (x_0 - \sqrt{A}y_0)^n. \qquad (49)$$

Adding and subtracting these equalities and dividing both sides of the relations by 2 or $2\sqrt{A}$, we obtain

$$x = x_n = \frac{1}{2}\left\{(x_0 + \sqrt{A}y_0)^n + (x_0 - \sqrt{A}y_0)^n\right\},$$

$$y = y_n = \frac{1}{2\sqrt{A}}\left\{(x_0 + \sqrt{A}y_0)^n - (x_0 - \sqrt{A}y_0)^n\right\}.$$

(50)

In other words, we obtain an explicit expression for an arbitrary solution $[x, y]$ for positive x and y. Each solution is obtained from these if we choose arbitrary signs for x_n and y_n.

For example, since, as we have seen above, the minimal solution for the equation $x^2 - 2y^2 = 1$ is $x = 3$, $y = 2$, then all positive solutions are contained in the equations

$$x_n = \frac{1}{2}[(3 + 2\sqrt{2})^n + (3 - 2\sqrt{2})^n],$$

$$y_n = \frac{1}{2\sqrt{2}}[(3 + 2\sqrt{2})^n - (3 - 2\sqrt{2})^n].$$

For $n = 1, 2, 3$ we obtain the solutions $[3, 2]$, $[17, 12]$, $[99, 70]$.

We note that, with increasing n, the numbers x_n and y_n grow at the same rate as the terms of a geometric progression with ratio $x_0 + \sqrt{A}y_0$ since in consequence of the equality

$$(x_0 + \sqrt{A}y_0)(x_0 - \sqrt{A}y_0) = 1$$

we may verify that

$$0 < x_0 - \sqrt{A}y_0 < 1,$$

and this means that $(x_0 - \sqrt{A}y_0)^n$ tends to 0 as n increases.

We notice now that if equation *(29)* has one non-trivial solution, in other words, a solution with $y \neq 0$, then there is a minimal solution of the equation, and then all of its solutions can be obtained from formulas *(50)*. The question of the existence of a non-trivial solution of the equation for an orbitrary integer A with $\sqrt{A}$ irrational we have left open; but now we proceed to its discussion.

5. The General Case of Equations of Second Degree with Two Unknowns

We shall show in this section that when A is an arbitrary positive integer with irrational $\sqrt{A}$ the equation

$$x^2 - Ay^2 = 1 \qquad (51)$$

always has a non-trivial solution; in other words, there exists a pair of integers x_0 and y_0, $x_0, y_0 \neq 0$, which satisfy the equation. First of all, we show how to expand an arbitrary positive number into a continued fraction. The continued fraction expansion of $\sqrt{2}$, given in the preceding section, depended upon a special property of this square root.

Let α be an arbitrary positive number. Then there is always an integer which is less than or equal to α and greater than $\alpha - 1$. Such an integer is called the *integral part* of α and is designated by $[\alpha]$. The difference between α and its integral part is called the *fractional part* of α and designated by $\{\alpha\}$. From the definitions of the integral and fractional parts of α immediately follows the relation

$$\alpha - [\alpha] = \{\alpha\}$$

or

$$\alpha = [\alpha] + \{\alpha\} \qquad (52)$$

between them. Since the fractional part is the difference between the number and the greatest integer not exceeding the number, this fractional part is always less than unity and is non-negative. For example, the integral part of $\frac{27}{5}$ is 5 and its fractional part is $\frac{2}{5}$; the integral part of $\sqrt{2}$ is 1 and its fractional part is $\sqrt{2} - 1$; the integral part of $\sqrt[3]{52}$ is 3 and its fractional part is $\sqrt[3]{52} - 3$; and so forth.

These notions of integral and fractional parts of a positive number may be used in expanding the number into a continued fraction. We put

$$[\alpha] = q_1, \quad \{\alpha\} = \frac{1}{\alpha_1} .$$

Then

$$\alpha = q_1 + \frac{1}{\alpha_1} . \tag{53}$$

Since $\{\alpha\}$ is always less than unity, α_1 is always larger than unity. If α were itself an integer, then the fractional part would be zero and the equation would be $\alpha = q_1$. This special case is excluded in our considerations since we confine ourselves to the expansions of irrational numbers. Thus we may require that α_1 be a positive number which is greater than unity. This number α_1 we treat in the same way as α and write the equality

$$\alpha_1 = q_2 + \frac{1}{\alpha_2} , \quad q_2 = [\alpha_1] , \quad \frac{1}{\alpha_2} = \{\alpha_1\} .$$

Continuing this process, we obtain the series of equalities

$$\alpha = q_1 + \frac{1}{\alpha_1} , \quad q_1 = [\alpha] ,$$

$$\alpha_1 = q_2 + \frac{1}{\alpha_2} , \quad q_2 = [\alpha_1] ,$$

$$\alpha_2 = q_3 + \frac{1}{\alpha_3} , \quad q_3 = [\alpha_2] , \tag{54}$$

$$\cdot \quad \cdot \quad \cdot$$

$$\alpha_{n-1} = q_n + \frac{1}{\alpha_n} , \quad q_n = [\alpha_{n-1}] ,$$

$$\cdot \quad \cdot \quad \cdot \quad \cdot \quad \cdot \quad \cdot \quad \cdot \quad \cdot \quad \cdot \quad \cdot \quad \cdot$$

This process of forming the integers $q_1, q_2, \ldots$, in the case when α is a rational number is easily seen to yield the same results as the corresponding process based on the Euclidean algorithm [see equations (6)]. The process must break off after a finite number of steps for α rational. For α irrational the process continues indefinitely. Indeed, if for some n, α_n were an integer, then α_{n-1} would be rational, which in its turn would cause α_{n-2}

to be rational, and so forth, finally giving α_1 rational. From *(54)* we obtain the fraction

$$\alpha = q_1 + \cfrac{1}{q_2 + \cfrac{1}{q_3 + \cfrac{\ddots}{\quad + \cfrac{1}{q_n + \cfrac{1}{\alpha_n}}}}} \tag{55}$$

which, since n may be taken as large as desired, may be written in the form of an infinite continued fraction

$$\alpha = q_1 + \cfrac{1}{q_2 + \cfrac{1}{q_3 + \cfrac{\ddots}{\quad + \cfrac{1}{q_n + \cfrac{}{\ddots}}}}}$$

As we have already noted in Section 4, the relation *(8)* for the convergents remains true in our case since it did not depend on the finiteness or non-finiteness of the fractions. From *(8)*, as we have already seen, for the even convergents, we have *(25)*. This inequality *(25)* will again provide the basis for the proof of the existence of solutions of equation *(51)*, but the proof will be more complicated than in the case when $A = 2$. For further information on the theory of continued fractions, see *Continued Fractions* by A. Y. Khinchin. *

THEOREM III. *For an arbitrary positive integer A with irrational $\sqrt{A}$, the equation (51)*

$$x^2 - Ay^2 = 1$$

has a non-trivial solution $[x_0, y_0]$, $x_0 > 0$, $y_0 > 0$.

---

* See footnote on page 21 [*Tr.*].

Proof: Because of certain complications in the proof of the existence of a solution to equation *(51)*, we break this proof into a series of steps. The first step is to show the existence of a positive integer k for which the equation

$$x^2 - Ay^2 = k \qquad (56)$$

has an infinite number of positive integral solutions x and y. Indeed, let us consider the binomial $x^2 - Ay^2$ and substitute for x and y the numerator and denominator respectively of the even convergents of the continued fraction expansion of the irrational number $\alpha = \sqrt{A}$. Then.

$$z_{2n} = P_{2n}^2 - AQ_{2n}^2 = (P_{2n} - \alpha Q_{2n})(P_{2n} + \alpha Q_{2n}). \qquad (57)$$

But because

$$0 < P_{2n} - \alpha Q_{2n} < \frac{1}{Q_{2n+1}},$$

it immediately follows that

$$0 < P_{2n} + \alpha Q_{2n} = 2\alpha Q_{2n} + P_{2n} - \alpha Q_{2n} < 2\alpha Q_{2n} + \frac{1}{Q_{2n+1}}.$$

We shall make use of these last two inequalities for the estimation of z_{2n}. Replacing, with the help of these inequalities, both factors on the right side of equation *(57)* increases the magnitude, and we obtain the inequality

$$0 < z_{2n} < \frac{1}{Q_{2n+1}}\Big(2\alpha Q_{2n} + \frac{1}{Q_{2n+1}}\Big) < 2\alpha + 1, \qquad (58)$$

since Q_{2n} is smaller than Q_{2n+1}. Substituting P_{2n} and Q_{2n} in place of x and y respectively in the binomial

$$z = x^2 - Ay^2$$

gives a positive integral value for z. Thus all the numbers $z_2, z_4, \ldots, z_{2n}, \ldots$, are positive integers which do not exceed the fixed number $2\alpha + 1$. But, since $\alpha = \sqrt{A}$ is irrational, the

continued fraction is infinite, and this means there are infinitely many pairs P_{2n} and Q_{2n}. The number of distinct positive integers among $z_2, z_4, \ldots, z_{2n}, \ldots$, is finite since between 1 and the fixed number $2\alpha + 1$, which does not depend upon n, there are at most $[2\alpha + 1]$ integers. The infinite sequence $z_2, z_4, \ldots,$ $z_{2n}, \ldots$, consists then of some sequence of the integers 1, 2, 3, $\ldots, [2\alpha + 1]$, in which not necessarily all of the integers 1, 2, 3, $\ldots, [2\alpha + 1]$, actually occur. Since the infinite sequence $z_2, z_4, \ldots, z_{2n}, \ldots$, has only a finite number of distinct terms, there must be a number k $(1 \leq k \leq [2\alpha + 1])$ which is repeated in the sequence infinitely many times. In other words, among the pairs $[P_2, Q_2], [P_4, Q_4], \ldots, [P_{2n}, Q_{2n}], \ldots$, there are infinitely many for which the quantity $z = x^2 - Ay^2$ takes the value k when one substitutes them in place of x and y. Enumerate these pairs which are solutions of *(56)* for the given k and designate them by $[u_1, v_1], [u_2, v_2], \ldots, [u_n, v_n], \ldots$.
We then have

$$u_n^2 - Av_n^2 = k. \tag{59}$$

We note that the sequence of pairs $[u_1, v_1], [u_2, v_2], \ldots, [u_n, v_n],$ $\ldots$, is a subsequence of the pairs of numerators and denominators of the even convergents of the continued fraction expansion of α. If we could assert that $k = 1$ we would already have proved that *(51)* has infinitely many solutions in integers. Since we may not make this assertion, we suppose $k > 1$ (in the contrary case, when $k = 1$, all is already proved) and proceed to the second step of our proof. We shall now show that among the pairs $[u_1, v_1], \ldots, [u_n, v_n], \ldots$, there are infinitely many pairs giving the same remainder upon division by k; in other words, there exists a pair of non-negative numbers p and q, less than k, for which infinitely many of the pairs $[u_1, v_1], \ldots,$ $[u_n, v_n], \ldots$, will verify the equalities

$$u_n = a_n k + p, \quad v_n = b_n k + q, \tag{60}$$

where a_n and b_n are the integral parts of $\dfrac{u_n}{k}$ and $\dfrac{v_n}{k}$ respectively and p and q are the remainders. Indeed, if we divide u_n and v_n by the integer k, $k > 1$, we obtain expressions of the form *(60)*, where the remainders from the division will be, as always, between 1 and $k - 1$. Since the remainders from the division of the number u_n by k must be only the numbers 0, 1, 2, . . . , $k - 1$, and in the same way the remainders from the division of the number v_n by k must also be only the numbers 0, 1, 2, . . . , $k - 1$, then the number of pairs of remainders possible when one divides u_n and v_n by k will be $k \cdot k = k^2$. This is clear also from the fact that every pair $[u_n, v_n]$ corresponds to a pair of remainders $[p_n, q_n]$, where p_n and q_n may each take on at most k distinct values and therefore the pair at most k^2 values. Thus every pair of integers $[u_n, v_n]$ corresponds to a pair of remainders $[p_n, q_n]$ under division by k. But the number of distinct pairs of remainders is finite, not exceeding k^2, and the number of pairs $[u_n, v_n]$ is infinite. This means that in the series of pairs $[p_1, q_1], [p_2, q_2], . . . , [p_n, q_n], . . .$, there are only finitely many distinct pairs and therefore that some one pair is repeated an infinite number of times. Designating such a pair of remainders by $[p, q]$, we obtain the existence of infinitely many pairs $[u_n, v_n]$ for which equality *(60)* is valid. Since not all pairs $[u_n, v_n]$ satisfy equations *(60)* for the definite pair p and q whose existence we have just proved, we again enumerate those pairs $[u_n, v_n]$ which do satisfy *(60)*. We designate these pairs by $[R_n, S_n]$. Thus the infinite sequence of pairs $[R_1, S_1]$, $[R_2, S_2], . . . , [R_n, S_n], . . .$, is a subsequence of the pairs $[u_n, v_n]$ which in turn is a subsequence of the pairs of numerators and denominators of the even convergents of the continued fraction for α. Each pair of this sequence satisfies *(59)* and at the same time gives the same remainder p and q upon division by k. Now that we have established the existence of infinitely many pairs of positive integers R_n and S_n, we may proceed to the third and fourth steps of our proof.

We note, first of all, that the pair $[R_n, S_n]$, being the numerator and denominator pair of a convergent of a continued fraction, must be a relative prime pair, that is, R_n and S_n have no common divisors. Indeed, if in the relation *(24)* we replace k by $2k$ and put $\delta_{2k} = \dfrac{P_{2k}}{Q_{2k}}$, $\delta_{2k-1} = \dfrac{P_{2k-1}}{Q_{2k-1}}$, then from the equality

$$\frac{P_{2k}}{Q_{2k}} - \frac{P_{2k-1}}{Q_{2k-1}} = \frac{1}{Q_{2k}Q_{2k-1}},$$

multiplying both sides by $Q_{2k}Q_{2k-1}$, we obtain the equation

$$P_{2k}Q_{2k-1} - Q_{2k}P_{2k-1} = 1. \tag{61}$$

This relation between the integers P_{2k}, Q_{2k}, P_{2k-1}, Q_{2k-1} shows that if P_{2k} and Q_{2k} have a common divisor greater than 1, then the left-hand side is also divisible by this common divisor. But on the right of this equality, we have 1 which is not divisible by any such number greater than unity. This proves that the numerator and denominator of a convergent of a continued fraction are relatively prime. Therefore the numbers R_n and S_n are relatively prime. From relation *(7)* also follows immediately that

$$Q_2 < Q_4 < \ldots < Q_{2n} < \ldots .$$

Because R_n and S_n are relatively prime and the fact that the numbers $S_1, S_2, \ldots, S_n, \ldots$, (contained among the numbers Q_{2n} which are all distinct) are distinct, it follows that in the infinite series of fractions

$$\frac{R_1}{S_1}, \frac{R_2}{S_2}, \ldots, \frac{R_n}{S_n}, \ldots,$$

no two are equal. We write two equations, following from the definitions of R_n and S_n

$$R_1^2 - AS_1^2 = (R_1 - \alpha S_1)(R_1 + \alpha S_1) = k \tag{62}$$

and

$$R_2^2 - AS_2^2 = (R_2 - \alpha S_2)(R_2 + \alpha S_2) = k, \tag{63}$$

where, as before, $\alpha = \sqrt{A}$.

Further we have

$$(R_1 - \alpha S_1)(R_2 + \alpha S_2) = R_1 R_2 - A S_1 S_2 + \alpha(R_1 S_2 - S_1 R_2), \quad (64)$$

since $\alpha^2 = A$ and similarly

$$(R_1 + \alpha S_1)(R_2 - \alpha S_2) = R_1 R_2 - A S_1 S_2 - \alpha(R_1 S_2 - S_1 R_2). \quad (65)$$

But R_n and S_n upon division by k give remainders which are independent of n. Consequently, in virtue of relation *(60)*,

$$R_n = c_n k + p, \quad S_n = d_n k + q. \quad (66)$$

Therefore, with the help of simple transformations and replacements, we obtain the equalities

$$\begin{aligned}
R_1 R_2 - A S_1 S_2 &= R_1(c_2 k + p) - A S_1(d_2 k + q) \\
&= R_1[(c_2 - c_1)k + c_1 k + p] - A S_1[(d_2 - d_1)k + d_1 k + q] \\
&= R_1[(c_2 - c_1)k + R_1] - A S_1[(d_2 - d_1)k + S_1] \qquad (67) \\
&= k[R_1(c_2 - c_1) - A S_1(d_2 - d_1)] + R_1^2 - A S_1^2 \\
&= k[R_1(c_2 - c_1) - A S_1(d_2 - d_1) + 1] = k x_1,
\end{aligned}$$

where x_1 is an integer, since $R_1^2 - A S_1^2 = k$. Similarly

$$\begin{aligned}
R_1 S_2 - S_1 R_2 &= R_1[(d_2 - d_1)k + d_1 k + q] \\
&\qquad\qquad - S_1[(c_2 - c_1)k + c_1 k + p] \qquad (68) \\
&= R_1[(d_2 - d_1)k + S_1] - S_1[(c_2 - c_1)k + R_1] \\
&= k[R_1(d_2 - d_1) - S_1(c_2 - c_1)] = k y_1,
\end{aligned}$$

where y_1 is again an integer. It is necessary to verify that y_1 is not equal to zero. Indeed, if $y_1 = 0$, then

$$k y_1 = R_1 S_2 - R_2 S_1 = 0,$$

whence

$$\frac{R_1}{S_1} = \frac{R_2}{S_2}.$$

This last equality is not possible since we have established that all the fractions $\frac{R_n}{S_n}$ are distinct from each other. Equalities (67) and (68) prove that

$$(R_1 - \alpha S_1)(R_2 + \alpha S_2) = kx_1 + \alpha ky_1 = k(x_1 + \alpha y_1) \qquad (69)$$

and

$$(R_1 + \alpha S_1)(R_2 - \alpha S_2) = kx_1 - \alpha ky_1 = k(x_1 - \alpha y_1) \qquad (70)$$

Multiplying equations (62) and (63) and making use of (69) and (70) yields

$$
\begin{aligned}
k^2 &= (R_1^2 - AS_1^2)(R_2^2 - AS_2^2) \\
&= (R_1 - \alpha S_1)(R_2 + \alpha S_2)(R_1 + \alpha S_1)(R_2 - \alpha S_2) \\
&= k^2(x_1 + \alpha y_1)(x_1 - \alpha y_1) = k^2(x_1^2 - Ay_1^2).
\end{aligned}
\qquad (71)
$$

Canceling k^2 we finally obtain

$$x_1^2 - Ay_1^2 = 1. \qquad (72)$$

But y_1 is not equal to zero, and therefore x_1 may not be zero. In the contrary case, the left side would be a negative number and the right unity. Thus, even supposing k not equal to 1, we have found two non-zero numbers x_1 and y_1 which satisfy equation (51).

This completes the theory of equations of type (51), since we know that each equation with A integral, $A > 0$, $\sqrt{A}$ irrational always has a solution and with the help of the minimal solution, the existence of which is proved at the same time, we are able to construct all of its solutions.

In practice the smallest solution is found by selecting suitable x_0 and y_0.

We have therefore completely examined the case $A > 0$ and $\alpha = \sqrt{A}$ irrational in the equation

$$x^2 - Ay^2 = 1.$$

If $A > 0$ and $\alpha = \sqrt{A}$ is an integer, then this equation may be written in the form

$$x^2 - \alpha^2 y^2 = (x + \alpha y)(x - \alpha y) = 1,$$

and since α is an integer if x_0 and y_0 are integers satisfying the equation, it is necessary that the separate equations

$$x_0 + \alpha y_0 = 1, \quad x_0 - \alpha y_0 = 1$$

or the equations

$$x_0 + \alpha y_0 = -1, \quad x_0 - \alpha y_0 = -1$$

be satisfied since the product of two integers may be equal to unity when and only when each of them equals 1 or each of them equals -1. These two systems of two equations in two unknowns x_0 and y_0 have only the trivial solutions: $x_0 = 1$, $y_0 = 0; x = -1, y_1 = 0$. Thus, *equation (51) with A equal to the square of an integer has only the trivial solutions $x_0 = \pm 1$, $y_0 = 0$. In the same way equation (51) has only the trivial solutions when A is integral and negative* (for $A = -1$ also the symmetric trivial solutions $x_0 = 0, y_0 = \pm 1$ satisfy the equation).

We consider now the equation

$$x^2 - Ay^2 = C, \tag{73}$$

where A is positive and integral, C is an integer, and $\alpha = \sqrt{A}$ is an irrational number. We have already seen that for $C = 1$ this equation always has infinitely many solutions in integers x and y. For arbitrary C and A such an equation may not have any solution.

EXAMPLE. We shall show that the *equation*

$$x^2 - 3y^2 = -1 \tag{74}$$

is not solvable in integers x and y. We note, first of all, that the square of an odd number under division by 8 always gives a remainder of 1. Indeed, since any odd number a may be written in the form $a = 2N + 1$ where N is an integer, then

$$a^2 = (2N+1)^2 = 4N^2 + 4N + 1 = 4N(N+1) + 1 = 8M + 1, \quad (75)$$

where M is an integer because either N or $N + 1$ must be even. Further, if $[x_0, y_0]$ is a solution of equation *(74)*, then x_0 and y_0 may not both be even or both odd. If x_0 and y_0 were both even or both odd, then $x_0^2 - 3y_0^2$ would be even and not be equal to -1. Similarly, if x_0 were odd and y_0 even, then x_0^2 would give a remainder of 1 upon division by 4 but $-3y_0^2$ would be divisible by 4. Therefore $x_0^2 - 3y_0^2$ would give a remainder of 1 upon division by 4. This is impossible since under division by 4 the right side trivially gives the remainder -1 or $3 = 4 - 1$. Finally, if x_0 is even and y_0 odd, then x_0^2 is divisible by 4, and $-3y_0^2$ on the basis of equation *(75)* may be written in the form

$$-3y_0^2 = -3(8M + 1) = -24M - 3 = 4(-6M - 1) + 1$$

and therefore under division by 4 gives a remainder of 1. Thus $x_0^2 - 3y_0^2$ when divided by 4 must again give a remainder of 1, which, as we have already seen, is impossible. Therefore, there do not exist integers x_0 and y_0 which satisfy equation *(74)*.

We do not concern ourselves about the proper conditions to impose upon C and A in order that equation *(73)* will have a solution. This question is difficult and is solved with the help of the general theory of quadratic irrationalities in the algebraic theory of numbers. We do concern ourselves with the case where equation *(73)* has a non-trivial solution. As before, we call a solution $[x', y']$ non trivial if $x', y' \neq 0$. Thus let equation *(73)* have the non-trivial solution $[x', y']$; in other words, let

$$x'^2 - Ay'^2 = C. \quad (76)$$

We consider for the same A the equation

$$x^2 - Ay^2 = 1. \quad (77)$$

This equation has infinitely many solutions in integers for $A > 0$ and $\alpha = \sqrt{A}$ irrational. Any one of its solutions $[\bar{x}, \bar{y}]$ is given by

$$\bar{x} = \pm x_n, \quad \bar{y} = \pm y_n,$$

where x_n and y_n are defined by formulas *(50)*. Since $[\bar{x}, \bar{y}]$ is a solution of *(77)*, then

$$\bar{x}^2 - A\bar{y}^2 = (\bar{x} + \alpha\bar{y})(\bar{x} - \alpha\bar{y}) = 1.$$

Equation *(76)*, in its turn, may be written in the form

$$(x' + \alpha y')(x' - \alpha y') = C.$$

Multiplying these last two equalities, we obtain

$$(x' + \alpha y')(\bar{x} + \alpha\bar{y})(x' - \alpha y')(\bar{x} - \alpha\bar{y}) = C. \qquad (78)$$

But

$$(x' + \alpha y')(\bar{x} + \alpha\bar{y}) = x'\bar{x} + Ay'\bar{y} + \alpha(x'\bar{y} + y'\bar{x})$$

and in exactly the same way

$$(x' - \alpha y')(\bar{x} - \alpha\bar{y}) = x'\bar{x} + Ay'\bar{y} - \alpha(x'\bar{y} + y'\bar{x}).$$

Making use of these two equations, we may write equation *(78)* in the form

$$[x'\bar{x} + Ay'\bar{y} + \alpha(x'\bar{y} + y'\bar{x})]\,[x'\bar{x} + Ay'\bar{y} - \alpha(x'\bar{y} + y'\bar{x})] = C$$

or in the form

$$(x'\bar{x} + Ay'\bar{y})^2 - A(x'\bar{y} + y'\bar{x})^2 = C.$$

By this we have shown that if $[x', y']$ is a solution of equation *(73)*, then this equation will be satisfied also by the pair of numbers $[x, y]$:

$$x = x'\bar{x} + Ay'\bar{y}, \quad y = x'\bar{y} + y'\bar{x}, \qquad (79)$$

where $[\bar{x}, \bar{y}]$ is any solution of equation *(77)*. Thus we have proved that *if equation (73) has one solution, then it has infinitely many*.

It is impossible, of course, to assert that equations *(79)* give all solutions of equation *(73)*. In the theory of algebraic numbers, it is proved that all integral solutions of equation *(73)* may be obtained from equations *(79)* by using for $[x', y']$ one of a finite number of such solutions of equation *(73)*, this number being dependent upon A and C. The equation *(73)* may not have more than finitely many solutions when A is negative or the square of an integer. This simply demonstrated assertion we leave to the reader to prove. The solution of the most general equation of second degree in two unknowns, equations of the form

$$Ax^2 + Bxy + Cy^2 + Dx + Ey + F = 0, \qquad (80)$$

where A, B, C, D, E, and F are integral, reduces by means of substitutions for the variables to the solution of an equation of the form *(73)* with positive or negative A. Therefore the character of the solutions, if they exist, is the same as that of equations of type *(73)*. Summing up the above discussion, we we may now say that *equations of the second degree in two unknowns of type (80) may not have integral solutions, may have only finitely many solutions, or finally, may have infinitely many such solutions; moreover these solutions we take from a finite number of generalized geometric progressions given by equations (79)*. When the behavior and character of solutions in integers of equations of second degree with two unknowns are compared with the behavior of the solutions of equations of the first degree, one very essential feature is indicated. Namely, solutions of first degree equations, if they exist, form arithmetic progressions, whereas solutions of second degree equations, if there are infinitely many, are taken from a finite number of generalized geometric progressions. In other words, in the case of the second degree equation, pairs of integers which may be solutions of the equation are more thinly distributed than are solutions of an equation of first degree. This is not accidental. It will be shown that equations in two unknowns of degree higher than two, generally speaking, may have only finitely many solutions. Exceptions to this rule are extremely rare.

6. Equations in Two Unknowns of Degree Higher Than the Second

Equations in two unknowns of degree higher than the second almost always, with rare exceptions, have only a finite number of integral solutions x and y. We consider, first of all, the equation

$$a_0x^n + a_1x^{n-1}y + a_2x^{n-2}y^2 + \ldots + a_ny^n = c, \qquad (81)$$

where n is an integer greater than two and all the numbers $a_0, a_1, a_2, \ldots, a_n, c$ are integers.

As was proved at the beginning of the twentieth century by A. Thue, *such an equation has only a finite number of solutions in integers x and y with possible exception in the cases when the left homogeneous part of this equation is a power of a homogeneous binomial of first degree or trinomial of second degree.* In this last case our equation will have one of the two forms

$$(ax + by)^n = c_0, \quad (ax^2 + bxy + cy^2)^n = c_0$$

and by the same token reduces to an equation of the first or second degree since, if a solution exists, c_0 must be the nth power of an integer. We shall not explain fully the Thue method because of its complexity. We thus restrict ourselves to certain explanatory remarks, giving an indication of the character of the proof of the finiteness of the number of solutions of equation *(81)*.*

---

* The literature on this subject is collected, for example, in the survey article by A. O. Gelfond, "The Approximation of Algebraic Numbers by Algebraic Numbers and the Theory of Transcendental Numbers," *Uspehi Math. Sci.*, 4, 4(32), 1949, p. 19. (This paper appears in English as Translation Number 65 of The American Mathematical Society. See also *Transcendental and Algebraic Numbers* by A. O. Gelfond, Dover Pub., 1960 [*Tr.*].)

We divide both sides of *(81)* by y^n. Our equation then takes the form

$$a_0 \left(\frac{x}{y}\right)^n + a_1 \left(\frac{x}{y}\right)^{n-1} + \ldots + a_{n-1}\left(\frac{x}{y}\right) + a_n = \frac{c}{y^n}. \quad (82)$$

For simplicity of the exposition let us suppose not only that all roots of the equation

$$a_0 z^n + a_1 z^{n-1} + \ldots + a_{n-1}z + a_n = 0 \quad (83)$$

are different and that $a_0 a_n \neq 0$, but also that the roots of this equation do not satisfy any equation of lower degree with integral coefficients. This condition is basic in our problem.

In higher algebra it is proved that each algebraic equation has at least one root. From this it is easily seen that a polynomial is divisible by $z - \alpha$ if α is a root of the polynomial. It follows then that each polynomial may be represented in the form of a product

$$a_0 z^n + a_1 z^{n-1} + \ldots + a_n = a_0(z - \alpha_1)(z - \alpha_2)\ldots(z - \alpha_n), \quad (84)$$

where $\alpha_1, \alpha_2, \ldots, \alpha_n$ are the n roots of the given polynomial. Using this expression of a polynomial in the form of a product, we may write equation *(82)* in the form

$$a_0\left(\frac{x}{y} - \alpha_1\right)\left(\frac{x}{y} - \alpha_2\right)\ldots\left(\frac{x}{y} - \alpha_n\right) = \frac{c}{y^n}. \quad (85)$$

Let us assume that there exists an infinite set of integral solutions $[x_k, y_k]$ of equation *(85)*. This implies that there exist solutions for which y_k has an absolute value as large as is desired. For if there were infinitely many pairs with bounded y_k, that is, all y_k have absolute values less than some definite number, and arbitrarily large x_k, then the left side of equation *(85)* could be made arbitrarily large while the right side would remain bounded—and this is impossible. We may suppose, then, that y_k is very large. Then the right side of equation *(85)* will be small, and this means that the left side is also very small. But the left side of the equation is a product of n factors, the $\frac{x_k}{y_k} - \alpha_m$, and the integer a_0, which is not less than 1. This means that the left side can be small only under the condition that some one of the differences

$$\left| \frac{x_k}{y_k} - \alpha_m \right|$$

is small. It is clear that this difference can be small only when α_m is real; in other words, we may not have $\alpha_m = a + bi$, $b \neq 0$. In the contrary case, the modulus of our difference could not be arbitrarily small since

$$\left| \frac{x_k}{y_k} - a - bi \right| = \sqrt{\left(\frac{x_k}{y_k} - a \right)^2 + b^2} > \left| b \right| .$$

Two of these differences on the left side of equation (85) may not be small simultaneously since

$$\left| \left(\frac{x_k}{y_k} - \alpha_m \right) - \left(\frac{x_k}{y_k} - \alpha_s \right) \right| = \left| \alpha_m - \alpha_s \right| \neq 0 \quad (86)$$

owing to the fact that the α_m are distinct. If one difference has absolute value smaller than $\frac{1}{2} \left| \alpha_m - \alpha_s \right|$ then in virtue of equation (86) the other will have absolute value larger than $\frac{1}{2} \left| \alpha_m - \alpha_s \right|$. This is a consequence of the fact that the absolute value of a sum does not exceed the sum of the absolute values. Since all of the numbers α_m are distinct, the smallest modulus $\left| \alpha_m - \alpha_s \right|$ of their differences will be larger than zero ($m \neq s$). If we designate this quantity by $2d$, then we shall obtain for sufficiently large y_k, which must occur since the y_k are unbounded,

$$\left| \frac{x_k}{y_k} - \alpha_m \right| < d$$

and

$$\left| \frac{x_k}{y_k} - \alpha s \right| > d, \quad s = 1, 2, \ldots, n, \quad s \neq m. \quad (87)$$

Then since the absolute value, or modulus, of a product is equal to the product of the absolute values, or moduli, of the factors, we have from equation *(85)*

$$|a_0| \left| \frac{x_k}{y_k} - \alpha_1 \right| \ldots \left| \frac{x_k}{y_k} - \alpha_{m-1} \right| \left| \frac{x_k}{y_k} - \alpha_m \right| \left| \frac{x_k}{y_k} - \alpha_{m+1} \right| \qquad (88)$$

$$\ldots \left| \frac{x_k}{y_k} - \alpha_n \right| = \frac{|c|}{|y_k|^n}.$$

If we replace in this equality each of the differences

$$\left| \frac{x_k}{y_k} - \alpha_s \right|, \qquad s \neq m,$$

by the smaller quantity d and replace $|a_0|$ by 1 ($|a_0|$ is not smaller than 1), then the left side of equation *(88)* is smaller than the right and we obtain the inequality

$$d^{n-1} \left| \frac{x_k}{y_k} - \alpha_m \right| < \frac{|c|}{|y_k|^n},$$

or

$$\left| \frac{x_k}{y_k} - \alpha_m \right| < \frac{c_1}{|y_k|^n}, \quad c_1 = \frac{|c|}{d^{n-1}}, \qquad (89)$$

where c_1 does not depend upon x_n and y_n. The number of α_m is no larger than n but there are infinitely many pairs $[x_k, y_k]$, each satisfying *(89)* for some m. Therefore there exists a definite m such that for the corresponding α_m the inequality *(89)* is correct for infinitely many pairs. In other words, if equation *(81)* has an infinite number of solutions in integers, then the algebraic equation *(83)* with integral coefficients has a root α for which for arbitrarily large q the inequality

$$\left| \alpha - \frac{p}{q} \right| < \frac{A}{q^n}, \qquad (90)$$

is true, where A is a constant independent of p and q, p and q

are integers, and n is the degree of the equation which α satisfies. If α were an arbitrary real number, it would be possible to choose it so that there are infinitely many integer solutions p, q to equation *(90)*. But in our case α is a root of an algebraic equation with integral coefficients. Such numbers are called algebraic and possess special properties. The *degree of an algebraic number* is the smallest degree of an algebraic equation with integral coefficients satisfied by the number.

Thue proved that for an algebraic number α of degree n the inequality

$$\left| \alpha - \frac{p}{q} \right| < \frac{1}{q^{\frac{n}{2}+1}}, \qquad n > 3, \qquad (91)$$

may have only finitely many solutions in integers p and q.* But if $n \geq 3$ the right side of inequality *(90)* (for sufficiently large q) is smaller than the right side of inequality *(91)* since $n > \frac{n}{2} + 1$. Thus, if *(91)* can have only finitely many solutions in integers p and q, then *(90)* more certainly has only a finite number of integral solutions. This means that equation *(81)* may have only finitely many solutions in integers when no root of equation *(83)* is a root of an algebraic equation with integral coefficients and of degree smaller than n. For $n = 2$, as may be easily established, inequality *(90)* may actually have infinitely many solutions in integers p and q for certain A. The theorem of Thue has been strengthened considerably. It is necessary only to remark that the method of proof of this theorem does not provide a way of finding an upper bound to the magnitudes of the solutions; in other words, the bounds of the possible magnitudes $|x|$ and $|y|$ depend upon the coefficients $a_0, a_1, \ldots, a_n$, and c. This question also remains open today. Although Thue's method does not afford a way of determining bounds on the magnitude of solutions, it does

---

* For an exposition of recent improvements in this result (by C. L. Siegel and K. F. Roth), see *Topics in Number Theory*, vol. 2, by W. J. LeVeque [*Tr.*].

make it possible to find bounds (though crude ones) on the number of solutions of equation *(83)*. For certain classes of equations of type *(83)*, these bounds may be quite precise. For example, the Soviet mathematician B. N. Delauney proved that the equation

$$ax^3 + y^3 = 1$$

for *a* integral may have, besides the trivial solution $x = 0, y = 1$, no more than one solution in integers *x* and *y*. Besides this, he proved that the equation

$$ax^3 + bx^2y + cxy^2 + dy^3 = 1$$

may have no more than five solutions in integers *x* and *y* for integral *a*, *b*, *c*, *d*.

Let $P(x, y)$ be an arbitrary polynomial in *x* and *y* with integral coefficients; in other words,

$$P(x, y) = \Sigma A_{ks} x^k y^s,$$

where A_{ks} are integers. We shall say that this is an *irreducible polynomial* if it cannot be written as the product of two other polynomials with integral coefficients each of which is not a constant.

By special and very complicated methods, C. L. Siegel proved that the equation

$$P(x, y) = 0,$$

where $P(x, y)$ is an irreducible polynomial in *x* and *y* of degree higher than the second (i.e., there appear in it terms of the form $A_{ks} x^k y^s$ where $k + s > 2$), may have infinitely many solutions in integers *x* and *y* only when there exist numbers a_n, $a_{n-1}, \ldots, a_0, a_{-1}, \ldots, a_{-n}$ and $b_n, b_{n-1}, \ldots, b_0, b_{-1}, \ldots, b_{-n}$ such that when substituting for *x* and *y* the expressions

$$x = a_n t^n + a_{n-1} t^{n-1} + \ldots + a_0 + \frac{a_{-1}}{t} + \ldots + \frac{a_{-n}}{t^n},$$

$$y = b_n t^n + b_{n-1} t^{n-1} + \ldots + b_0 + \frac{b_{-1}}{t} + \ldots + \frac{b_{-n}}{t^n}$$

in our equation we obtain the identity

$$P(x, y) \equiv 0,$$

with respect to t. Here n is some integer.

7. Algebraic Equations of Degree Higher Than the Second with Three Unknowns and an Exponential Equation

For equations with two unknowns, we are able to answer the question of the existence of a finite or infinite number of integral solutions. For equations of degree higher than the second, with more than two unknowns, we may answer this question only in very special cases, and for these equations even less can be said about the determination of all integral solutions. We conclude with the so-called Fermat conjecture as an example.

The famous French mathematician Pierre Fermat made the assertion that the equation

$$x^n + y^n = z^n \qquad (92)$$

has no positive integral solutions for integral $n \geq 3$ (the case $xyz = 0$ is excluded by the positivity of x, y, z). Despite Fermat's assertion that he had a proof (probably by the method of descent, about which a word will be said below), this proof has never been found. Moreover, when the mathematician E. Kummer attempted to find a proof and even thought at one time that he had one, he discovered that a proposition true for ordinary integers becomes untrue for more complicated number domains which are encountered in a natural way in the investigation of the Fermat problem. The denial of this proposition for the algebraic integers, that is, the roots of algebraic equations with integral coefficients and coefficient of highest power 1, consists of the fact that they may not be factorizable in a unique way into prime (non-factorable in their turn) factors which are algebraic integers. The ordinary integers may be factored into primes in a unique way. For example, $6 = 2 \cdot 3$, and no other factorization is possible within the set of ordinary integers. Consider now the set of all algebraic integers of the form $m + n\sqrt{-5}$, where m and n are ordinary integers. It is easy to see that the sum and product of two such numbers are again numbers of the set. A set of numbers possessing the property that it contains any sum and product of numbers in the set is called a ring. The ring defined above contains the numbers 2, 3, $1 + \sqrt{-5}$, $1 - \sqrt{-5}$. Each of the numbers in this ring can be easily shown to be prime; that is, they are not representable as the product of two ($\neq 1$) numbers of the ring. But

$$6 = 2 \cdot 3 = (1 + \sqrt{-5})(1 - \sqrt{-5});$$

in other words, the number 6 does not have a unique prime factorization in our ring. This non-unique factorization into primes is possible in other more complicated rings of algebraic integers. Discovering this, Kummer convinced himself that his proof of the Fermat conjecture was false. In order to overcome the difficulties connected with the non-unique factorization, Kummer constructed the theory of ideals which plays so large a role in present-day algebra and number theory. But even with the help of this new theory, Kummer was not able to prove the Fermat conjecture. He proved it only for those n which are divisible by one of the so-called regular prime numbers. Although we cannot stop to discuss the concept of regular primes, we do want to observe that at the present time it is not known whether there exists a finite or infinite number of them.

Today the Fermat conjecture has been proved for many n, in particular for any n divisible by a prime less than 100. The Fermat conjecture has played a large role in the development of mathematics, thanks to the opening of the theory of ideals in the attempt to prove it. But in this connection, it is necessary to note that, in a quite different way and for other reasons, this theory was constructed by the famous Russian mathematician E. I. Solotarov, who died in the bloom of his scientific work. At the present time, a special proof of the Fermat conjecture, especially one based on considerations of the divisibility of numbers, may have only a sporting interest. Of course, if this proof is obtained by new and fruitful methods, then its value may be very great, owing to the significance of the methods. It is necessary to remark that those attempts of amateur mathematicians to prove the Fermat conjecture by quite elementary methods are doomed to failure. Elementary considerations based on the divisibility of numbers were exploited by Kummer, and continued examination of these considerations by outstanding mathematicians has up to now yielded nothing essential.

We proceed here to the proof of the Fermat theorem in the case $n = 4$, since the method of descent, on which the proof is constructed, is very interesting.

THEOREM IV. *The Fermat equation*

$$x^4 + y^4 = z^4 \qquad (93)$$

has no integral solutions for x, y, and z, $xyz \neq 0$.

Proof: We will prove an even stronger theorem, namely, that the equation

$$x^4 + y^4 = z^2 \qquad (94)$$

has no integral solutions for $x, y,$ and z, $xyz \neq 0$. From this theorem it immediately follows that equation *(93)* has no solution. If equation *(94)* has a solution in integers x, y, z all different from 0, then it is possible to suppose these integers are relatively prime in pairs. Indeed, if there is a solution in which the numbers x and y have a greatest common divisor $d > 1$ then

$$x = dx_1, \quad y = dy_1$$

where $(x_1, y_1) = 1$. Dividing both sides of equation *(94)* by d^4, we have

$$x_1^4 + y_1^4 = \left(\frac{z}{d^2}\right)^2 = z_1^2. \tag{95}$$

But x_1 and y_1 are integers, which implies $z_1 = \frac{z}{d^2}$ is also an integer. If z_1 and y_1 had a common divisor $k > 1$ then, in virtue of equation *(95)*, x_1^2 would be divisible by k, which implies that x_1 and k would not be relatively prime. Thus we have shown that if there exists a solution of equation *(94)* in integers, distinct from zero, then there exists such a solution in integers distinct from zero and relatively prime. Therefore it is sufficient for us to prove that equation *(94)* has no solution in integers which are distinct from zero and relatively prime in pairs. In the course of this proof, when we say that equation *(94)* has a solution, we shall presuppose that it has a solution in positive integers, relatively prime in pairs.

In Section 3 we have shown that all solutions of equation *(12)*

$$x^2 + y^2 = z^2 \tag{96}$$

in positive integers, relatively prime in pairs, are defined by formulas *(18)* and have the form

$$x = uv, \quad y = \frac{u^2 - v^2}{2}, \quad z = \frac{u^2 + v^2}{2}, \tag{97}$$

where u and v are two odd, relatively prime, positive integers.

We give a somewhat different form to the formulas *(97)* which define all solutions of equation *(96)*. Since u and v are odd numbers, then putting

$$\frac{u + v}{2} = a, \quad \frac{u - v}{2} = b, \tag{98}$$

we obtain u and v from the equalities

$$u = a + b, \quad v = a - b, \tag{99}$$

where a and b are integers of different parity. Equalities *(98)* and *(99)* show that each pair of relatively prime odd integers u and v corresponds to a pair of relatively prime integers a and b of different parity, and that each pair of relatively prime integers a and b of different parity corresponds to a pair of relatively prime odd integers u and v. Therefore making the replacements u and v for a and b in equations *(97)*, we find that all triples of positive pairwise relatively prime integers x, y, z (x odd) which are solutions of equation *(96)* are determined by the formulas

$$x = a^2 - b^2, \quad y = 2ab, \quad z = a^2 + b^2, \tag{100}$$

where a and b are two relatively prime integers of different parity subject to the condition $x > 0$. These formulas show that x and y are of different parity. If equation *(94)* has the solution x_0, y_0, z_0 then

$$[x_0^2]^2 + [y_0^2]^2 = z_0^2;$$

in other words, the triple of numbers (x_0^2, y_0^2, z_0) is a solution of equation *(96)*. But then there must exist two numbers a and b, $a > b$, relatively prime and of different parity such that

$$x_0^2 = a^2 - b^2, \quad y_0^2 = 2ab, \quad z_0 = a^2 + b^2. \tag{101}$$

For definiteness we have assumed that x_0 is odd and y_0 even. The opposite assumption does not change the argument since we would then just replace x_0 by y_0 and conversely. But we already know [see equality *(75)*] that the square of an odd number leaves a remainder of 1 when divided by 4. Therefore from the equality

$$x_0^2 = a^2 - b^2 \qquad (102)$$

it follows that a is odd and b is even. In the contrary case, the left side of this equality upon division by 4 gives a remainder of 1 and the right, since we assume a even and b odd, -1. Since a is odd and $(a, b) = 1$, then $(a, 2b) = 1$. But then from the equality

$$y_0^2 = 2ba,$$

it follows that

$$a = t^2, \quad 2b = s^2, \qquad (103)$$

where t and s are integers. But from the relation *(102)*, it follows that $[x_0, b, a]$ is a solution of equation *(96)*. This implies

$$x_0 = m^2 - n^2, \quad b = 2mn, \quad a = m^2 + n^2,$$

where m and n are certain relatively prime numbers of different parity. From *(103)* we have

$$mn = \frac{b}{2} = \left(\frac{s}{2}\right)^2$$

whence, since m and n are relatively prime, it follows that

$$m = p^2, \quad n = q^2, \qquad (104)$$

where p and q are integers distinct from zero. Since $a = t^2$ and $a = m^2 + n^2$, then

$$q^4 + p^4 = t^2. \qquad (105)$$

But

$$z_0 = a^2 + b^2 > a^2.$$

Therefore

$$0 < t = \sqrt{a} < \sqrt[4]{z_0} < z_0 \qquad (z_0 > 1). \qquad (106)$$

Putting $q = x_1$, $p = y_1$, and $t = z_1$ we see that if there exists a solution $[x_0, y_0, z_0]$, then there necessarily exists another solution $[x_1, y_1, z_1]$ where $0 < z_1 < z_0$. This process of obtaining solutions of equation *(94)* may be continued indefinitely and we obtain a sequence of solutions

$$[x_0, y_0, z_0], [x_1, y_1, z_1], \ldots, [x_n, y_n, z_n], \ldots,$$

in which the positive integers $z_0, z_1, z_2, \ldots, z_n, \ldots$, are monotone decreasing; in other words, the inequalities

$$z_0 > z_1 > z_2 > \ldots > z_n > \ldots$$

are valid. But positive integers cannot form an infinite monotone decreasing sequence since in such a sequence no term is greater than z_0. We arrive, therefore, at a contradiction when we assume that equation *(94)* has at least one solution in integers x, y, z, $xyz \neq 0$. This proves that equation *(94)* has no solution. Consequently equation *(93)* also has no solution in positive integers $[x, y, z]$ since in the contrary case a solution $[x, y, z]$ of equation *(93)* would give the solution $[x, y, z^2]$ of equation *(94)*.

The method of proof which we exploited, that is, the construction with the help of one solution of an infinite sequence of solutions with indefinitely decreasing positive integers z, is called the method of descent. As we have already noted, the use of this method of proof is hindered as long as the decomposition of the integers in the algebraic ring into prime factors of the same ring is non-unique. *

---

* For further information concerning the Fermat conjecture, see *The Fermat Conjecture* by A. Y. Khinchin. (Since this book is in Russian, see "Fermat's Last Theorem" by H. S. Vandiver, *Am. Math. Mo.*, *53*, 10 (1946), 555–578, and the supplementary note by the same author, *Am. Math. Mo.*, *60*, 3 (1953), 164–167 [*Tr.*].)

Observe that we have proved the absence of integral solutions not only of equation *(94)* but also of the equation

$$x^{4n} + y^{4n} = z^{2n}.$$

It is interesting to note that the equation

$$x^4 + y^2 = z^2$$

has an infinite number of positive integral solutions—for example, $x = 2, y = 3, z = 5$. The finding of the form of all positive integral solutions of this equation we leave to the reader.

We give one more example of the method of descent with a somewhat different course of reasoning.

EXAMPLE. *We prove that the equation*

$$x^4 + 2y^4 = z^2 \tag{107}$$

does not have a solution in integers x, y, z all different from zero. Suppose that equation *(107)* has a solution in positive integers $[x_0, y_0, z_0]$. We may suppose that these numbers are relatively prime since if there were a greatest common divisor $d > 1$, the numbers $\frac{x_0}{d}, \frac{y_0}{d}, \frac{z_0}{d^2}$ would be such a solution of equation *(107)*. The existence of a common divisor of two of them easily leads to the existence of a common divisor for all three. In addition, we suppose that z_0 is the least possible value of z for all solutions of equation *(107)* in positive integers. Since $[x_0, y_0, z_0]$ is a solution of equation *(107)*, then $[x_0^2, y_0^2, z_0]$ is a solution of the equation

$$x^2 + 2y^2 = z^2. \tag{108}$$

Using formulas *(19')* of Section 3, giving all positive integral solutions of *(108)*, we see that there exist positive integers a and b such that $(a, b) = 1$, a is odd, and that satisfy the equations

$$x_0^2 = \pm(a^2 - 2b^2), \quad y_0^2 = 2ab, \quad z_0 = a^2 + 2b^2. \tag{109}$$

From the equality of $y_0^2 = 2ab$ it follows that b must be even since y_0 is even, y_0^2 is divisible by 4, and a is odd. Since $\frac{b}{2}$ and a are relatively prime, then from

$$\left(\frac{y_0}{2}\right)^2 = a\left(\frac{b}{2}\right)$$

it follows immediately that

$$a = m^2, \quad \frac{b}{2} = n^2,$$

where m and n are positive integers and $(m, 2n) = 1$. But from equation *(109)* follows

$$x_0^2 = \pm(a^2 - 2b^2) = \pm\left[a^2 - 8\left(\frac{b}{2}\right)^2\right], \qquad (110)$$

where x_0 and a are odd. We have already seen that the square of an odd number leaves a remainder of 1 when divided by 4. Therefore the left side of equation *(110)* under division by 4 leaves a remainder of 1. Also $a^2 - 8\left(\frac{b}{2}\right)^2$ gives a remainder of 1 under division by 4. This means that the bracketed term on the right side of equation *(110)* must appear only with a plus. Now equation *(110)* may be written in the form

$$x_0^2 = m^4 - 8n^4$$

or in the form

$$x_0^2 + 2(2n^2)^2 = (m^2)^2, \qquad (111)$$

where x_0, n, and m are positive relatively prime integers. This means that the numbers x_0, $2n^2$, m^2 form a solution of equation *(108)*, where x_0, $2n^2$, m^2 are relatively prime. Therefore, again in virtue of formulas *(19′)* of Section 3, we find integers p and q, p odd, $(p, q) = 1$, such that

$$2n^2 = 2pq, \quad m^2 = p^2 + 2q^2, \quad x_0 = \pm(p^2 - 2q^2). \qquad (112)$$

But since $(p, q) = 1$ and $n^2 = pq$, then

$$p = s^2, \quad q = r^2,$$

where s and r are relatively prime integers. Whence finally the relation

$$s^4 + 2r^4 = m^2 \tag{113}$$

follows. This shows that the numbers s, r, m form a solution of equation *(107)*. But from the above equalities

$$z_0 = a^2 + 2b^2, \qquad a = m^2$$

it follows that $z_0 > m$. Thus from a solution $[x_0, y_0, z_0]$ we have found another solution $[x, r, m]$ where $0 < m < z_0$. This contradicts our assumption that $[x_0, y_0, z_0]$ was a solution for which z_0 was as small as possible. Thus, assuming the existence of a solution of equation *(107)*, we have arrived at a contradiction and have proved that this equation is insolvable in integers distinct from zero.

We leave it to the reader to prove that the equations

$$x^4 + 4y^4 = z^2, \quad x^4 - y^4 = z^2,$$
$$x^4 - y^4 = 2z^2, \quad x^4 - 4y^4 = z^2$$

are insolvable in positive integers.

In conclusion, we make some remarks about exponential equations. *The equation*

$$a^x + b^y = c^z, \tag{114}$$

where a, b, and c are integers, not 0 or powers of 2, may have no more than a finite number of solutions in integers x, y, z. This same assertion remains true, with a small additional condition, when a, b, and c are arbitrary algebraic numbers. Moreover, the equation

$$A\alpha_1^{x_1} \ldots \alpha_n^{x_n} + B\beta_1^{y_1} \ldots \beta_m^{y_m} + C\gamma_1^{z_1} \ldots \gamma_p^{z_p} = 0, \tag{115}$$

where A, B, C, $ABC \neq 0$, are integers, $\alpha_1, \ldots, \alpha_n$, $\beta_1, \ldots, \beta_m$, $\gamma_1, \ldots, \gamma_p$ are integers and the numbers α, β, γ,

$$\alpha = \alpha_1 \ldots \alpha_n, \quad \beta = \beta_1 \ldots \beta_m, \quad \gamma = \gamma_1 \ldots \gamma_p,$$

are relatively prime, may have only a finite number of solutions in integers $x_1, \ldots, x_n, y_1, \ldots, y_m, z_1, \ldots, z_p$. This assertion also generalizes in the case where A, B, C and α_i, β_k, γ_s are algebraic.[*] Equations of type *(115)* and their generalizations are of great interest since in the theory of algebraic numbers it is proved that every algebraic equation of type *(81)* corresponds to a certain exponential equation of type *(115)* where each solution of *(81)* corresponds to a solution of *(115)* in integers. Such a correspondence extends also to equations of more general type than *(81)* and *(115)*.

---

[*] See the survey paper by A. O. Gelfond cited on page 46.

INDEX

A CATALOG OF SELECTED
DOVER BOOKS
IN SCIENCE AND MATHEMATICS

Mathematics–Algebra and Calculus

VECTOR CALCULUS, Peter Baxandall and Hans Liebeck. This introductory text offers a rigorous, comprehensive treatment. Classical theorems of vector calculus are amply illustrated with figures, worked examples, physical applications, and exercises with hints and answers. 1986 edition. 560pp. 5 3/8 x 8 1/2.   0-486-46620-5

ADVANCED CALCULUS: An Introduction to Classical Analysis, Louis Brand. A course in analysis that focuses on the functions of a real variable, this text introduces the basic concepts in their simplest setting and illustrates its teachings with numerous examples, theorems, and proofs. 1955 edition. 592pp. 5 3/8 x 8 1/2.   0-486-44548-8

ADVANCED CALCULUS, Avner Friedman. Intended for students who have already completed a one-year course in elementary calculus, this two-part treatment advances from functions of one variable to those of several variables. Solutions. 1971 edition. 432pp. 5 3/8 x 8 1/2.   0-486-45795-8

METHODS OF MATHEMATICS APPLIED TO CALCULUS, PROBABILITY, AND STATISTICS, Richard W. Hamming. This 4-part treatment begins with algebra and analytic geometry and proceeds to an exploration of the calculus of algebraic functions and transcendental functions and applications. 1985 edition. Includes 310 figures and 18 tables. 880pp. 6 1/2 x 9 1/4.   0-486-43945-3

BASIC ALGEBRA I: Second Edition, Nathan Jacobson. A classic text and standard reference for a generation, this volume covers all undergraduate algebra topics, including groups, rings, modules, Galois theory, polynomials, linear algebra, and associative algebra. 1985 edition. 528pp. 6 1/8 x 9 1/4.   0-486-47189-6

BASIC ALGEBRA II: Second Edition, Nathan Jacobson. This classic text and standard reference comprises all subjects of a first-year graduate-level course, including in-depth coverage of groups and polynomials and extensive use of categories and functors. 1989 edition. 704pp. 6 1/8 x 9 1/4.   0-486-47187-X

CALCULUS: An Intuitive and Physical Approach (Second Edition), Morris Kline. Application-oriented introduction relates the subject as closely as possible to science with explorations of the derivative; differentiation and integration of the powers of x; theorems on differentiation, antidifferentiation; the chain rule; trigonometric functions; more. Examples. 1967 edition. 960pp. 6 1/2 x 9 1/4.   0-486-40453-6

ABSTRACT ALGEBRA AND SOLUTION BY RADICALS, John E. Maxfield and Margaret W. Maxfield. Accessible advanced undergraduate-level text starts with groups, rings, fields, and polynomials and advances to Galois theory, radicals and roots of unity, and solution by radicals. Numerous examples, illustrations, exercises, appendixes. 1971 edition. 224pp. 6 1/8 x 9 1/4.   0-486-47723-1

AN INTRODUCTION TO THE THEORY OF LINEAR SPACES, Georgi E. Shilov. Translated by Richard A. Silverman. Introductory treatment offers a clear exposition of algebra, geometry, and analysis as parts of an integrated whole rather than separate subjects. Numerous examples illustrate many different fields, and problems include hints or answers. 1961 edition. 320pp. 5 3/8 x 8 1/2.   0-486-63070-6

LINEAR ALGEBRA, Georgi E. Shilov. Covers determinants, linear spaces, systems of linear equations, linear functions of a vector argument, coordinate transformations, the canonical form of the matrix of a linear operator, bilinear and quadratic forms, and more. 387pp. 5 3/8 x 8 1/2.   0-486-63518-X

Browse over 9,000 books at www.doverpublications.com

Mathematics–Probability and Statistics

BASIC PROBABILITY THEORY, Robert B. Ash. This text emphasizes the probabilistic way of thinking, rather than measure-theoretic concepts. Geared toward advanced undergraduates and graduate students, it features solutions to some of the problems. 1970 edition. 352pp. 5 3/8 x 8 1/2. 0-486-46628-0

PRINCIPLES OF STATISTICS, M. G. Bulmer. Concise description of classical statistics, from basic dice probabilities to modern regression analysis. Equal stress on theory and applications. Moderate difficulty; only basic calculus required. Includes problems with answers. 252pp. 5 5/8 x 8 1/4. 0-486-63760-3

OUTLINE OF BASIC STATISTICS: Dictionary and Formulas, John E. Freund and Frank J. Williams. Handy guide includes a 70-page outline of essential statistical formulas covering grouped and ungrouped data, finite populations, probability, and more, plus over 1,000 clear, concise definitions of statistical terms. 1966 edition. 208pp. 5 3/8 x 8 1/2. 0-486-47769-X

GOOD THINKING: The Foundations of Probability and Its Applications, Irving J. Good. This in-depth treatment of probability theory by a famous British statistician explores Keynesian principles and surveys such topics as Bayesian rationality, corroboration, hypothesis testing, and mathematical tools for induction and simplicity. 1983 edition. 352pp. 5 3/8 x 8 1/2. 0-486-47438-0

INTRODUCTION TO PROBABILITY THEORY WITH CONTEMPORARY APPLICATIONS, Lester L. Helms. Extensive discussions and clear examples, written in plain language, expose students to the rules and methods of probability. Exercises foster problem-solving skills, and all problems feature step-by-step solutions. 1997 edition. 368pp. 6 1/2 x 9 1/4. 0-486-47418-6

CHANCE, LUCK, AND STATISTICS, Horace C. Levinson. In simple, non-technical language, this volume explores the fundamentals governing chance and applies them to sports, government, and business. "Clear and lively ... remarkably accurate." – *Scientific Monthly.* 384pp. 5 3/8 x 8 1/2. 0-486-41997-5

FIFTY CHALLENGING PROBLEMS IN PROBABILITY WITH SOLUTIONS, Frederick Mosteller. Remarkable puzzlers, graded in difficulty, illustrate elementary and advanced aspects of probability. These problems were selected for originality, general interest, or because they demonstrate valuable techniques. Also includes detailed solutions. 88pp. 5 3/8 x 8 1/2. 0-486-65355-2

EXPERIMENTAL STATISTICS, Mary Gibbons Natrella. A handbook for those seeking engineering information and quantitative data for designing, developing, constructing, and testing equipment. Covers the planning of experiments, the analyzing of extreme-value data; and more. 1966 edition. Index. Includes 52 figures and 76 tables. 560pp. 8 3/8 x 11. 0-486-43937-2

STOCHASTIC MODELING: Analysis and Simulation, Barry L. Nelson. Coherent introduction to techniques also offers a guide to the mathematical, numerical, and simulation tools of systems analysis. Includes formulation of models, analysis, and interpretation of results. 1995 edition. 336pp. 6 1/8 x 9 1/4. 0-486-47770-3

INTRODUCTION TO BIOSTATISTICS: Second Edition, Robert R. Sokal and F. James Rohlf. Suitable for undergraduates with a minimal background in mathematics, this introduction ranges from descriptive statistics to fundamental distributions and the testing of hypotheses. Includes numerous worked-out problems and examples. 1987 edition. 384pp. 6 1/8 x 9 1/4. 0-486-46961-1

Browse over 9,000 books at www.doverpublications.com

Mathematics–Geometry and Topology

PROBLEMS AND SOLUTIONS IN EUCLIDEAN GEOMETRY, M. N. Aref and William Wernick. Based on classical principles, this book is intended for a second course in Euclidean geometry and can be used as a refresher. More than 200 problems include hints and solutions. 1968 edition. 272pp. 5 3/8 x 8 1/2. 0-486-47720-7

TOPOLOGY OF 3-MANIFOLDS AND RELATED TOPICS, Edited by M. K. Fort, Jr. With a New Introduction by Daniel Silver. Summaries and full reports from a 1961 conference discuss decompositions and subsets of 3-space; n-manifolds; knot theory; the Poincaré conjecture; and periodic maps and isotopies. Familiarity with algebraic topology required. 1962 edition. 272pp. 6 1/8 x 9 1/4. 0-486-47753-3

POINT SET TOPOLOGY, Steven A. Gaal. Suitable for a complete course in topology, this text also functions as a self-contained treatment for independent study. Additional enrichment materials make it equally valuable as a reference. 1964 edition. 336pp. 5 3/8 x 8 1/2. 0-486-47222-1

INVITATION TO GEOMETRY, Z. A. Melzak. Intended for students of many different backgrounds with only a modest knowledge of mathematics, this text features self-contained chapters that can be adapted to several types of geometry courses. 1983 edition. 240pp. 5 3/8 x 8 1/2. 0-486-46626-4

TOPOLOGY AND GEOMETRY FOR PHYSICISTS, Charles Nash and Siddhartha Sen. Written by physicists for physics students, this text assumes no detailed background in topology or geometry. Topics include differential forms, homotopy, homology, cohomology, fiber bundles, connection and covariant derivatives, and Morse theory. 1983 edition. 320pp. 5 3/8 x 8 1/2. 0-486-47852-1

BEYOND GEOMETRY: Classic Papers from Riemann to Einstein, Edited with an Introduction and Notes by Peter Pesic. This is the only English-language collection of these 8 accessible essays. They trace seminal ideas about the foundations of geometry that led to Einstein's general theory of relativity. 224pp. 6 1/8 x 9 1/4. 0-486-45350-2

GEOMETRY FROM EUCLID TO KNOTS, Saul Stahl. This text provides a historical perspective on plane geometry and covers non-neutral Euclidean geometry, circles and regular polygons, projective geometry, symmetries, inversions, informal topology, and more. Includes 1,000 practice problems. Solutions available. 2003 edition. 480pp. 6 1/8 x 9 1/4. 0-486-47459-3

TOPOLOGICAL VECTOR SPACES, DISTRIBUTIONS AND KERNELS, François Trèves. Extending beyond the boundaries of Hilbert and Banach space theory, this text focuses on key aspects of functional analysis, particularly in regard to solving partial differential equations. 1967 edition. 592pp. 5 3/8 x 8 1/2.
0-486-45352-9

INTRODUCTION TO PROJECTIVE GEOMETRY, C. R. Wylie, Jr. This introductory volume offers strong reinforcement for its teachings, with detailed examples and numerous theorems, proofs, and exercises, plus complete answers to all odd-numbered end-of-chapter problems. 1970 edition. 576pp. 6 1/8 x 9 1/4. 0-486-46895-X

FOUNDATIONS OF GEOMETRY, C. R. Wylie, Jr. Geared toward students preparing to teach high school mathematics, this text explores the principles of Euclidean and non-Euclidean geometry and covers both generalities and specifics of the axiomatic method. 1964 edition. 352pp. 6 x 9. 0-486-47214-0

Browse over 9,000 books at www.doverpublications.com

Mathematics-History

THE WORKS OF ARCHIMEDES, Archimedes. Translated by Sir Thomas Heath. Complete works of ancient geometer feature such topics as the famous problems of the ratio of the areas of a cylinder and an inscribed sphere; the properties of conoids, spheroids, and spirals; more. 326pp. 5 3/8 x 8 1/2.   0-486-42084-1

THE HISTORICAL ROOTS OF ELEMENTARY MATHEMATICS, Lucas N. H. Bunt, Phillip S. Jones, and Jack D. Bedient. Exciting, hands-on approach to understanding fundamental underpinnings of modern arithmetic, algebra, geometry and number systems examines their origins in early Egyptian, Babylonian, and Greek sources. 336pp. 5 3/8 x 8 1/2.   0-486-25563-8

THE THIRTEEN BOOKS OF EUCLID'S ELEMENTS, Euclid. Contains complete English text of all 13 books of the Elements plus critical apparatus analyzing each definition, postulate, and proposition in great detail. Covers textual and linguistic matters; mathematical analyses of Euclid's ideas; classical, medieval, Renaissance and modern commentators; refutations, supports, extrapolations, reinterpretations and historical notes. 995 figures. Total of 1,425pp. All books 5 3/8 x 8 1/2.

Vol. I: 443pp.   0-486-60088-2
Vol. II: 464pp.   0-486-60089-0
Vol. III: 546pp.   0-486-60090-4

A HISTORY OF GREEK MATHEMATICS, Sir Thomas Heath. This authoritative two-volume set that covers the essentials of mathematics and features every landmark innovation and every important figure, including Euclid, Apollonius, and others. 5 3/8 x 8 1/2.

Vol. I: 461pp.   0-486-24073-8
Vol. II: 597pp.   0-486-24074-6

A MANUAL OF GREEK MATHEMATICS, Sir Thomas L. Heath. This concise but thorough history encompasses the enduring contributions of the ancient Greek mathematicians whose works form the basis of most modern mathematics. Discusses Pythagorean arithmetic, Plato, Euclid, more. 1931 edition. 576pp. 5 3/8 x 8 1/2.

0-486-43231-9

CHINESE MATHEMATICS IN THE THIRTEENTH CENTURY, Ulrich Libbrecht. An exploration of the 13th-century mathematician Ch'in, this fascinating book combines what is known of the mathematician's life with a history of his only extant work, the Shu-shu chiu-chang. 1973 edition. 592pp. 5 3/8 x 8 1/2.

0-486-44619-0

PHILOSOPHY OF MATHEMATICS AND DEDUCTIVE STRUCTURE IN EUCLID'S ELEMENTS, Ian Mueller. This text provides an understanding of the classical Greek conception of mathematics as expressed in Euclid's Elements. It focuses on philosophical, foundational, and logical questions and features helpful appendixes. 400pp. 6 1/2 x 9 1/4.   0-486-45300-6

BEYOND GEOMETRY: Classic Papers from Riemann to Einstein, Edited with an Introduction and Notes by Peter Pesic. This is the only English-language collection of these 8 accessible essays. They trace seminal ideas about the foundations of geometry that led to Einstein's general theory of relativity. 224pp. 6 1/8 x 9 1/4.   0-486-45350-2

HISTORY OF MATHEMATICS, David E. Smith. Two-volume history – from Egyptian papyri and medieval maps to modern graphs and diagrams. Non-technical chronological survey with thousands of biographical notes, critical evaluations, and contemporary opinions on over 1,100 mathematicians. 5 3/8 x 8 1/2.

Vol. I: 618pp.   0-486-20429-4
Vol. II: 736pp.   0-486-20430-8

Browse over 9,000 books at www.doverpublications.com